355.8
HUG

© 2005

# The Manhattan Project
## Big Science and the Atom Bomb

Jeff Hughes

**Revolutions in Science**
Series editor: Jon Turney

 **Columbia University Press**  New York

Columbia University Press
*Publishers Since 1893*
New York    Chichester, West Sussex

Copyright © 2002 Jeff Hughes
First published by Icon Books Ltd., Duxford

Library of Congress Cataloging-in-Publication Data
Hughes, Jeff (Jeff A.)
The Manhattan Project : big science and the atom bomb / Jeff Hughes.
p. cm.
Originally published: Duxford, Cambridge : Icon Books, 2002.
(Revolutions in science)
Includes bibliographical references.
ISBN 0–231–13152–6 (cloth : alk. paper)
1. Manhattan Project (U.S.) 2. Atomic bomb—United States—History.
I. Title.  II. Revolutions in science.
QC733.3.U5H84 2003
355.8′25119′0973—dc21    2003051620

∞

Columbia University Press books are printed on
permanent and durable acid-free paper.
Printed in the United States of America
c 10 9 8 7 6 5 4 3 2 1

# CONTENTS

# ILLUSTRATIONS

## Acknowledgements

Special thanks to all at City View for giving me space and peace to write, and to all at #170 for the music and support. Thanks also to Jon Agar and Jon Turney for helpful comments on an early draft, and to Simon Flynn for encouragement.

I am grateful to all the historians on whose work I have drawn, and have tried to indicate my debts in the section on 'Further Reading' at the end of the book. Thanks, too, to my students at the University of Manchester over the last few years, with whom I have enjoyed exploring the ideas that follow.

Above all, thanks to Lauren for friendship and love. For putting up with my obsession with books and bombs, I dedicate this to her.

# Chapter 1

# Introduction:
# Big Science and the Bomb

During the twentieth century, almost every aspect of science changed. Geographically, science spread from a few countries to many. Institutionally, it spread from universities and specialist organisations to find new homes in government, public and private industry and the military. Intellectually, its contours changed with the development of entirely new disciplines and the blurring of boundaries between old ones. Science has become an economic force and a key indicator of national development. It changed people's lives through the development or improvement of a wide range of products and processes, including materials, medicines, fuels, foods and communication techniques.

Perhaps the most remarkable change in twentieth century science, however, was its growth in scale, scope and cost. Consider two images. Figure 1 shows the physicist Ernest Rutherford in his laboratory at McGill University, Montreal, in the early 1900s. Rutherford was a leading researcher in the new field of

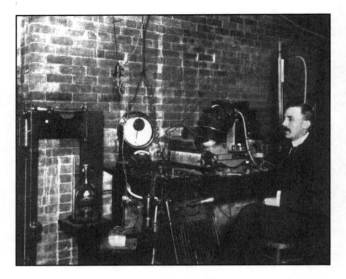

*Figure 1.* Small-scale science: Ernest Rutherford in his laboratory at McGill University, Montreal, in the early 1900s.

radioactivity. We see him at his laboratory bench, the sophisticated electrical equipment he used to measure the properties and behaviour of minute particles of matter set out in front of him. The apparatus was small-scale and relatively inexpensive, and could be comfortably accommodated on a tabletop and used effectively by one person. With the delicate traces from this apparatus, Rutherford explored the nature of the radiations emitted from radioactive substances, publishing papers under his name alone. With just a handful of people working in his own field, and just a thousand or so physicists in the world, a researcher like Rutherford could follow the scientific literature and know most of what was going on, both in his specialist field and in physics more broadly. He could make the decisions about the direction of his own research, and do that research pretty much alone.

Now look at Figure 2. It shows just one part of the detector called UA1 at CERN, the European

*Figure 2.* Big Science: part of the UA1 detector at CERN in the 1980s.

Organisation for Nuclear Research, in the 1980s. One of the most complex pieces of scientific equipment in the world, UA1 sat at the intersection of two high-energy beams of particles. After being speeded up in a particle accelerator, the particles smashed into each other, creating scores of smaller particles, whose tracks UA1 detected. From millions of such tracks, UA1 and the computers that controlled it sifted out just a few 'signature events' indicating the existence of new kinds of particles. Several storeys high and costing many millions of dollars, UA1 was constructed and operated by hundreds of scientists, engineers and technicians, with several different groups responsible for different parts of the apparatus and the larger experiment. In 1983, UA1 found evidence of the existence of new particles called the 'W' and 'Z'. Figure 3 shows the title page of the scientific paper announcing the discovery: it was authored by 138 people from twelve different laboratories in eight countries. The UA1 project was promoted and directed by an energetic and entrepreneurial Italian physicist, Carlo Rubbia, whose job was to manage the vast human, material and financial resources constituting the experiment, and to represent them to CERN and the outside world. For this, Rubbia shared the 1984 Nobel Physics Prize.

But UA1 was just a small part of CERN. Figure 4 indicates the size and scale of the whole laboratory today. As you can see, its scale is vast. By the late twentieth century, the size of experiments in high-

**EXPERIMENTAL OBSERVATION OF ISOLATED LARGE TRANSVERSE ENERGY ELECTRONS WITH ASSOCIATED MISSING ENERGY AT $\sqrt{s}$ = 540 GeV**

UA1 Collaboration, CERN, Geneva, Switzerland

G. ARNISON[j], A. ASTBURY[j], B. AUBERT[b], C. BACCI[i], G. BAUER[i], A. BÉZAGUET[d], R. BÖCK[d], T.J.V. BOWCOCK[f], M. CALVETTI[d], T. CARROLL[d], P. CATZ[b], P. CENNINI[d], S. CENTRO[d], F. CERADINI[d], S. CITTOLIN[d], D. CLINE[i], C. COCHET[k], J. COLAS[b], M. CORDEN[c], D. DALLMAN[d], M. DeBEER[k], M. DELLA NEGRA[b], M. DEMOULIN[d], D. DENEGRI[k], A. Di CIACCIO[i], D. DiBITONTO[d], L. DOBRZYNSKI[g], J.D. DOWELL[c], M. EDWARDS[c], K. EGGERT[a], E. EISENHANDLER[f], N. ELLIS[d], P. ERHARD[a], H. FAISSNER[a], G. FONTAINE[g], R. FREY[h], R. FRÜHWIRTH[l], J. GARVEY[c], S. GEER[g], C. GHESQUIÈRE[g], P. GHEZ[b], K.L. GIBONI[a], W.R. GIBSON[f], Y. GIRAUD-HÉRAUD[g], A. GIVERNAUD[k], A. GONIDEC[b], G. GRAYER[j], P. GUTIERREZ[h], T. HANSL-KOZANECKA[a], W.J. HAYNES[j], L.O. HERTZBERGER[2], C. HODGES[h], D. HOFFMANN[a], H. HOFFMANN[d], D.J. HOLTHUIZEN[2], R.J. HOMER[c], A. HONMA[f], W. JANK[d], G. JORAT[d], P.I.P. KALMUS[f], V. KARIMÄKI[e], R. KEELER[f], I. KENYON[c], A. KERNAN[h], R. KINNUNEN[e], H. KOWALSKI[d], W. KOZANECKI[h], D. KRYN[d], F. LACAVA[d], J.-P. LAUGIER[k], J.-P. LEES[b], H. LEHMANN[a], K. LEUCHS[a], A. LÉVÊQUE[k], D. LINGLIN[b], E. LOCCI[k], M. LORET[k], J.-J. MALOSSE[k], T. MARKIEWICZ[d], G. MAURIN[d], T. McMAHON[c], J.-P. MENDIBURU[g], M.-N. MINARD[b], M. MORICCA[i], H. MUIRHEAD[d], F. MULLER[d], A.K. NANDI[j], L. NAUMANN[d], A. NORTON[d], A. ORKIN-LECOURTOIS[g], L. PAOLUZI[i], G. PETRUCCI[d], G. PIANO MORTARI[i], M. PIMIÄ[e], A. PLACCI[d], E. RADERMACHER[a], J. RANSDELL[h], H. REITHLER[a], J.-P. REVOL[d], J. RICH[k], M. RIJSSENBEEK[d], C. ROBERTS[j], J. ROHLF[d], P. ROSSI[d], C. RUBBIA[d], B. SADOULET[d], G. SAJOT[g], G. SALVI[f], G. SALVINI[i], J. SASS[k], J. SAUDRAIX[k], A. SAVOY-NAVARRO[k], D. SCHINZEL[f], W. SCOTT[j], T.P. SHAH[j], M. SPIRO[k], J. STRAUSS[i], K. SUMOROK[c], F. SZONCSO[j], D. SMITH[h], C. TAO[d], G. THOMPSON[f], J. TIMMER[d], E. TSCHESLOG[a], J. TUOMINIEMI[e], S. Van der MEER[d], J.-P. VIALLE[d], J. VRANA[g], V. VUILLEMIN[d], H.D. WAHL[i], P. WATKINS[c], J. WILSON[c], Y.G. XIE[d], M. YVERT[b] and E. ZURFLUH[d]

*Figure 3.* The title page of one of the CERN papers announcing the discovery of the W and Z particles, illustrating the large-scale and international nature of the UA1 collaboration.

energy physics was measured in square miles, rather than table-tops. With this huge increase in size went a massive increase in organisational complexity. CERN employs thousands of people, with thousands more visiting each year to work there – not just on UA1 but on many other detectors and experiments. Such is the scale and complexity of these machines and institutions that entire communities of scientists can now spend their entire time working on one small part of an experiment. Much of the decision-making about what kinds of experiments should be carried out and how they should be organised is done by dozens

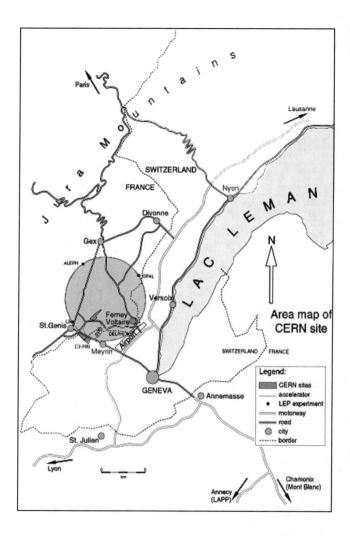

*Figure 4.* The geographical extent of contemporary Big Science: the site of the Large Hadron Collider being built at CERN, showing the path of the 27-kilometre tunnel.

of panels and committees. For example, in its range and complexity (with separate divisions for computing, personnel and so on), Fermilab, CERN's chief competitor in the USA, is more reminiscent of an industrial organisation than of the kind of enterprise we saw Rutherford engaged in. By the early 1990s, American physicists were planning an even larger particle accelerator, the Superconducting Super Collider (SSC). The SSC was to have been built in Texas, and would have dwarfed CERN (Figure 5). It was to cost several billion dollars – a substantial portion of the American science budget – and would include two underground experimental halls each the size of a football stadium, putting UA1 in the shade. Quite a difference from Rutherford alone in his lab!

How and why did science make this change from the small-scale, table-top science of Rutherford to the astonishing 'Big Science' of CERN and the planned SSC? From being an individual activity in which one person could carry out the whole of an experiment from planning to publication, how did science become a large-scale enterprise carried out by multi-disciplinary and multinational groups of hundreds of researchers working intensively within huge organisations? From being a relatively inexpensive activity carried out on table-tops in small laboratories, why did it come to demand massive institutions of its own, costing such enormous sums, often representing a significant fraction of national budgets? How, in short, did the Little Science of the first half of the

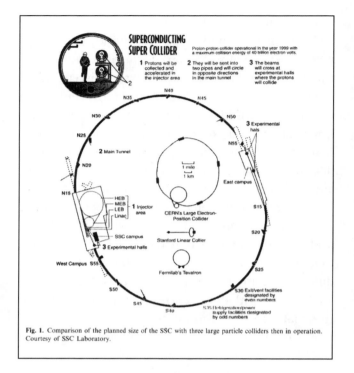

**Fig. 1.** Comparison of the planned size of the SSC with three large particle colliders then in operation. Courtesy of SSC Laboratory.

*Figure 5.* A 1990s comparison of the size of the projected Superconducting Super Collider with the existing Large Electron-Positron Collider at CERN.

twentieth century become the Big Science of the post-war period?

One answer often given to this question is the Manhattan Project: the mission by British and American scientists to develop nuclear weapons during World War Two. During the war, scientists and engineers contributed many developments that shaped the course of the conflict, including radar, penicillin, rockets and other new types of weaponry. But in particular, the Manhattan Project created new links between scientists and the state, and put science – especially physics – at the heart of national security. From 1943 to 1945 at huge sites across the United States, the Manhattan Project brought together thousands of scientists, engineers and technicians in secret to work on the design and production of atomic bombs. At its height, the Project employed 130,000 people and was equivalent in size to the entire American automobile industry. Its ultimate cost was $2 billion. All this effort was directed to one overriding goal: the production of nuclear weapons. Originally intended as a response to the threat of Hitler's acquiring nuclear weapons, they were ultimately dropped on Japan, destroying the cities of Hiroshima and Nagasaki, and hastening the end of the war.

The Manhattan Project was a turning point in the history of twentieth-century science. Nuclear weapons became key elements of post-war military strategy. As a Cold War developed between the West and the Soviet Union, science and scientists – especially physicists – became important national assets. Funding for military technical and scientific developments skyrocketed, and powerful networks of military, university and industrial scientists worked to develop new generations of weapons. Nuclear science in particular acquired a new prestige, and its practitioners were able to exploit state concerns about national security and nuclear weapons to develop ever-larger machines for exploring the structure of matter. Drawing on their experiences of organised large-scale research during the war, and often on wartime contacts, nuclear scientists planned big new machines and institutions both to carry on the research they had been doing before the war on the properties of matter and to explore the new possibilities of nuclear physics raised by the wartime project. In the United States and the Soviet Union, and to a lesser extent in Europe, they received lavish state funding, often through military organisations, for their work.

As discoveries began to flow from their machines and as they learned more and more about the structure of matter, the physicists began planning machines of ever-higher energies. Even by the 1950s these machines were becoming so expensive that smaller nations could scarcely afford them. In Europe, twelve nations combined to create CERN – the acronym originally stood for Conseil Européen pour la Recherche Nucléaire – in order to be able to compete effectively with the United States. Over the next forty years, while the superpowers competed with each other in an ever more expensive nuclear arms race, US and European physicists competed with each other in 'pure' science to achieve ever-larger machines and ever-higher energies, in the hope of discovering ever-smaller particles. By the 1970s, three major accelerator centres in the United States were contending with

CERN in Europe for the highest energy and the opportunity to make groundbreaking discoveries that would bring prestige and perhaps even Nobel Prizes for project leaders. Rubbia would win this race in 1983.

According to this version of events, Big Science was the logical consequence of the Manhattan Project and the status that nuclear weapons and nuclear energy gave to physics. But perhaps the story of Big Science is not quite as simple as that. Even at the height of its success, all was not happy in the expanding world of big machines, big experiments, big payrolls and big money. In the early 1960s, the American physicist Alvin Weinberg, director of the Oak Ridge National Laboratory (a former Manhattan Project laboratory) criticised the trend to ever-larger machines and ever more complex organisations in science. He coined the term 'Big Science' to describe the gigantism that he saw as coming to dominate science and especially scientific funding. Weinberg argued that Big Science was bad for science, because it encouraged scientists to spend money on large-scale research instead of developing well-thought-out experiments. He pointed out that at the rate Big Science was growing, it threatened to swallow up all scientific funding and ruin both science and the economy. He criticised the complex bureaucracies and media manipulation that were necessary to manage Big Science, and suggested that, in its obsession with machines and competition, Big Science had lost touch with human problems. Serious choices would soon have to be made between different kinds of science, he argued: scientists and science policy-makers would be faced with such decisions as whether to fund nuclear physics or oceanography, to put a man on Mars or to cure cancer.

In 1963, the American historian of science Derek de Solla Price took up Weinberg's argument, and planted the term Big Science firmly in the public mind with the title of his famous book: *Little Science, Big Science*. Price took a longer-term perspective than Weinberg on the development of Big Science. He argued that it was not simply a post-war phenomenon, 'an urgent public reaction to the first atomic explosion and the first national shocks of military missiles and satellites ... with historical roots no deeper in time than the Manhattan Project, Cape Canaveral rocketry, the discovery of penicillin, and the invention of radar and electronic computers'. For Price, Big Science was a relative term. He argued that large-scale science had been around for a much longer time, and that World War Two was only a perturbation in a longer-term dynamic in the growth of science. Certainly big, expensive instruments seem always to have been around in science in some shape or form. Similarly, many of the organisational characteristics attributed to Manhattan Project science were in place long before World War Two. Machines, hierarchical division of teamwork, goal-directed projects, close relationships between scientists, industry and the military: all existed well before 1939. Entrepreneurial scientists wanting to raise large sums of money for their work developed links with journalists in the inter-war years, and some laboratories even had unofficial 'press agents' who would ensure them good press coverage. 'Money, manpower, machines, media and the military' – the five Ms of Big Science – all seem to have been in place before the Manhattan Project and the bomb.

There are other criticisms of the idea of Big Science too. While we tend to think of Big Science as being physics and engineering projects centred on large institutions like particle accelerators, it need be neither. Think, for example, of the Human Genome Project (HGP): a huge, lavishly funded and goal-directed enterprise based on molecular biology, the HGP is spread across many institutions and involves small-scale research by many small groups of people. Oceanography, the Hubble Space Telescope and numerous other examples remind us that we have to be careful not to equate Big Science with just physics, and with just large institutions. Maybe the big historical picture of Big Science is not as simple as we think.

In this book, we will explore the nature of Big Science throughout the twentieth century and investigate its connections with the Manhattan Project and other military projects of World War Two. We shall see that the Manhattan Project did not cause a radical change in the development of science; rather, it accelerated developments already taking place. Focussing mainly on the USA and Europe, we shall see how the military and industry have gradually come to shape twentieth-century science more and more strongly. On our way, we will learn much about the connections between scientific organisation and scientific knowledge. We shall follow the progress of Big Science as it grew ever bigger, until by the 1980s some people were calling it 'Megascience'. And we shall ask whether in the last decade of the twentieth century, after the end of the Cold War, Big Science finally reached its end. The cancellation of the Superconducting Super Collider project in 1993 was a watershed in the history of Big Science. Coupled with the extensive growth of the biological sciences best evidenced by the Human Genome Project, the fate of the SSC suggests that we may be coming to the end of a particular way of doing science, and that at the beginning of the twenty-first century we are now facing a fundamental reorientation of scientific aims and values.

· Chapter 2 ·

# Long Before the Bomb: The Origins of Big Science

In defining Big Science, Derek de Solla Price pointed out that it was not solely a twentieth-century phenomenon. The scale and complexity of a science must obviously be seen in the context of its time. There have been numerous examples of large-scale or national science projects in the long history of science, including the observatories of Ulugh Beg in Samarkand as far back as the fifteenth century, of the Danish astronomer Tycho Brahe at Hven in the sixteenth and of Jai Singh in India in the seventeenth. The growth of interest in astronomy in the late eighteenth and early nineteenth centuries and the development of precision telescopy led William Parsons (later Lord Rosse) to start building large telescopes in Ireland, where his family had settled. In 1839 he built a 36-inch reflector, and followed it six years later with a 6-foot mirror weighing 4 tons, and with a 54-foot focal length. The mirrors were mounted in a wooden tube 7 feet in diameter and 58 feet long. The huge structure was mounted between two massive walls, 70 feet long and 50 feet high, to support and shield the telescope from the wind. The biggest telescope in the world when it was unveiled, it became known as the 'Leviathan of Parsonstown' (Figure 6). It is said that Dean Peacock of the Church of Ireland officially opened the telescope by walking through the tube wearing his top hat and carrying a raised umbrella over his head!

Within weeks of the Leviathan seeing first light, Rosse announced the discovery of the spiral structure of the Whirlpool galaxy in the constellation of Canes Venatici (The Hunting Dogs) – the first recognition of

*Figure 6.* Nineteenth-century Big Science: the 'Leviathan of Parsonstown', the telescope constructed by William Parsons in Ireland in the 1840s.

a spiral form in a nebula. The eminent English astronomer John Herschel said that the building of the great reflector was 'an achievement of such magnitude ... that I want words to express my admiration of it'. But the institutionalisation and growth of astronomy in the nineteenth century brought about changes in the practice of observation. For example, as the Royal Observatory, Greenwich, expanded from two staff in 1811 to 53 in 1900, George Airy (Astronomer Royal from 1835 to 1881) introduced a system in which young assistants were carefully trained and disciplined in their observations in order to manage differences between observers. Timekeeping practices and a division of labour were established to ensure efficient operation; indeed, the Greenwich regime displayed a 'factory mentality', closely linking it to the workings of the industrial society around it. As we shall see, this idea of *managing* significant numbers of researchers and operatives in a large institution within the framework of a single scientific enterprise was later to be important in Big Science.

Through the early twentieth century, astronomy continued to set the pace for the institutionalisation of large-scale science. In the 1880s, the Lick Observatory opened on Mt. Hamilton in Santa Clara County, California: the same area that, decades later, would become home to Silicon Valley. Funded by a large bequest, the project involved a huge construction effort, including the building of a road to the isolated hilltop site, chosen for its isolation from city lights. A 12-inch telescope was erected in a great dome atop a building 235 feet in circumference, at a total cost of some $70,000. It became the first permanently occupied mountaintop observatory in the world. Twenty years later, a 60-inch telescope was built at Mt. Wilson, above the Los Angeles basin in California. It was funded by a private foundation, the Carnegie Institution of Washington, and was then the world's largest telescope. Even as it was being built, its planners were designing an even larger, 100-inch instrument. Having faced significant technical challenges, such as the casting and shaping of the glass mirror, the 100-inch telescope was completed in 1917. The American astronomer Edwin Hubble (later to be commemorated in the Hubble Space Telescope, a Big Science project) used it to measure the distances and velocities of neighbouring galaxies, and with it showed that the universe is expanding – evidence for the cosmological Big Bang model.

Telescopes and observatories continued to grow in size and cost. But astronomy was not the only science to adopt large equipment requiring groups of workers and substantial funding. By the beginning of the twentieth century, a range of sciences were carried out either with extensive and expensive equipment or on a vast scale. At the Royal Institution in London, associated in the nineteenth century with Humphry Davy and Michael Faraday, for example, their successor James Dewar used the well-equipped laboratory to carry out research on low temperatures – the foundations of what we today would call cryogenics. At a time when the refrigeration industry was becoming important for the transportation of food over long distances, technologists and scientists became interested in ways of producing and using low temperatures. Dewar used pumps, compressors and other large-scale mechanical equipment to create temperatures within 15°C of absolute zero (–273°C) to liquefy gases – famously, in 1895, hydrogen. He invented a double-walled flask with a vacuum between the walls to hold his liquefied gases – and the Dewar flask is the basis of today's thermos flasks. The low temperatures he produced also allowed him to study the properties of solids under these unusual conditions. Dewar used students and two engineers to help with his experiments, and his laboratory in the basement of the Royal Institution looked more like a ship's engine room than a traditional chemical laboratory.

Dewar was not the only scientist wrestling with machines and nature to produce low temperatures. Across the Channel at Leiden in the Netherlands, the Dutch physicist Heike Kamerlingh Onnes also established a large laboratory devoted to low temperatures. A firm believer in 'knowledge through measurement', he gave much thought to the organisation of his laboratory and its work. His quest for accuracy was supported by the belief that 'a modern physical laboratory must be modelled upon astronomical lines'

– i.e. like the big observatories we discussed earlier. Kamerlingh Onnes used a staff of skilled assistants, who had trained in a school for instrument-makers and glassblowers funded by the Netherlands Education Agency. In 1885 he established a journal – *Communications from the Physical Laboratory at the University of Leiden* – to disseminate results from his laboratory; this eventually became the 'bible' of low-temperature physics. Kamerlingh Onnes' research was largely funded by the refrigeration industry, and as the size of its equipment increased in the early twentieth century, it came to look more and more like a factory, and became jokingly (but tellingly) known as the brewery. Racing with Dewar in London to achieve the next milestone in low-temperature physics, Kamerlingh Onnes and his associates liquefied helium for the first time in 1908, bringing researchers within 1.5°C of absolute zero. In 1911, in a series of experiments on electrical resistance at low temperatures, they also discovered that the resistance of mercury disappeared suddenly at about –269°C: this phenomenon would later come to be understood as superconductivity.

With its demands for large sums of money, complex instrumentation, organised labour and a close relationship with industry, low-temperature physics was not untypical of early twentieth-century physics. We find many of the same characteristics even if we think of what is usually taken as one of the 'purest' branches of science: atomic physics. Though we saw him earlier as a lone experimenter, from 1898 to 1907 at McGill University, Montreal, Ernest Rutherford developed a school of research in the new science of radioactivity. Money was critical to his success: William Macdonald, a tobacco dealer who hated smoking, donated what were, for the time, huge sums for the construction, fitting-out and maintenance of a physics laboratory, for apparatus, for employees' wages and for a research fund. Impressed by Rutherford's energy and commitment, he gave the young scientist more money to buy equipment and materials. In particular, one high-cost item was radium. Discovered by Pierre and Marie Curie in 1898, radium was an essential prerequisite for radioactivity research: a crucial but expensive source of sub-atomic particles for the study of matter. But it was also exceedingly expensive, for it was only manufactured in small quantities and much of it went for medical uses. All told, Macdonald seems to have spent of the order of £100,000 on physics at McGill (and his gifts to McGill as a whole totalled some £1 million). This lavish funding undoubtedly made possible Rutherford's groundbreaking work on radioactivity and atomic structure. In several ways, radium was to Rutherford what big particle accelerators were to later particle physicists: a *very* expensive scientific tool.

In addition to the purchase of materials like radium, Macdonald's support allowed the purchase of machinery such as a liquid air plant and electrical equipment to expedite Rutherford's research programme.

Moreover, the complex apparatus needed to manipulate radioactive materials and the relatively few laboratories that possessed substantial quantities of such materials meant that, like large accelerators, radium research tended to occur at fixed-site facilities. Just as with later particle accelerators, radioactivity researchers would have to travel to a radium laboratory to learn their craft and carry out experimental work. In order to accommodate the many such researchers who wished to work with him, Rutherford developed a system of project management that distributed tasks to individual researchers within his own larger intellectual research programme. He typically worked by collaboration between himself or other experienced workers and novitiates, and promoted co-publication. The most useful collaborations were often multi-disciplinary. It was with the English chemist Frederick Soddy, for example, that Rutherford worked out the theory of radioactive transformation – the idea that radioactive elements decay by emitting alpha-, beta- or gamma-rays, and in so doing transform themselves into nearby elements in the periodic table. Multiply-authored papers were a characteristic of Rutherford's laboratories – though usually with only two or three co-authors, rather than hundreds – and so in terms of workforce, too, we can see that Rutherford had a Big Science of his own.

Expensive equipment, ready access to sources of funding, an extensive workforce organised centrally and hierarchically, a complex division of labour and corresponding publication strategy, a sense of being in a race against other laboratories, entrepreneurialism and empire building – characteristics that would become very familiar fifty years later – were already part of Rutherford's physics in the early twentieth century. As Professor of Physics at Manchester from 1907 to 1919, Rutherford would continue to develop the same strategies. At Manchester, he inherited one of the best-equipped physics laboratories in the British Empire (opened only in 1900, it had cost about £30,000 – an enormous sum for the time). He quickly secured sources of radium and other radioactive materials and again began to build up a research group, either working himself with students or pairing students off with each other or with experienced workers. In his attempts to develop a way of counting sub-atomic particles, for example, he worked with the German physicist Hans Geiger, who had been studying with Schuster at Manchester. They developed both optical and electrical methods for counting particles – and Geiger would later develop this electrical method into the instrument that now bears his name, the Geiger counter. For the optical method, in which particles were counted by the flashes they made on a scintillating screen, Rutherford needed younger co-workers (with better eyes than his own), and had to discipline them carefully to ensure that they produced reliable, consistent results – much as we saw Airy do at the Greenwich Observatory.

After he won the Nobel Prize for Chemistry in 1908, Rutherford's prestige and influence increased still further. Students came from all over the world to train with him, so that his laboratory had a cosmopolitan and active atmosphere. He continued to accumulate supplies of radioactive materials, allowing him to carry out new kinds of experiments and obtain new effects – often ones which no other worker could obtain, so that he and his laboratory were able to dominate the field of radioactivity. We shall see later how such command of expensive resources can be an important factor in scientific success. Rutherford also began to exercise considerable political influence in the world of science. His expertise got him involved in international discussions about radioactivity standards, and he was invited to elite international meetings of physicists. As the price of radium continued to soar, he advised widely on matters radioactive (for example, to hospitals), and used the media to good effect to publicise the field. In many ways, he was the Carlo Rubbia of his day.

With his younger co-workers, Rutherford also continued to experiment on the atom. One series of experiments involved firing energetic alpha-particles from a radioactive substance at thin sheets of gold foil. Most of the particles passed straight through the foil, but a small number bounced back in the direction from which they were fired. This astonishing result led Rutherford in 1911 to postulate the existence of the *nucleus*: a tiny electrically charged core containing most of the mass of the atom. It was repulsion by the intense nuclear charge that had caused the bombarding particles to bounce back. Though this theory was one among several competing ideas about the structure of the atom, it became increasingly accepted after Rutherford showed in 1917 that the nuclei of elements could sometimes be broken up – that pieces corresponding to the simplest nucleus, hydrogen, could sometimes be chipped off larger nuclei. Rutherford called these hydrogen nuclei *protons*. Most physicists then came to see the atom as being made up of a nucleus containing a number of protons corresponding to the number of the element in the periodic table, and a number of electrons circling the nucleus in orbits, rather like a solar system. Though the term 'nuclear physics' came later, a very few physicists now started to wonder about the structure of the nucleus itself.

## · CHAPTER 3 ·

# SCIENCE, THE MILITARY
# AND INDUSTRY:
# THE GREAT WAR AND AFTER

By the time he did this work, most of Rutherford's energies were directed elsewhere. After the outbreak of war in Europe in 1914, senior British scientists volunteered themselves and their laboratories for the conflict. In what would become known as the Great War, all the belligerents undertook military research and development, organising committees and mobilising resources to produce scientific weapons and counter-weapons which might be of use in the war. Infamously, British and German chemists produced chemical weapons such as mustard gas for use on the battlefields – it has been estimated that some 5,500 scientists worked on chemical weapons alone during the conflict. The appalling casualties produced by these scientifically-produced yet hard-to-control weapons, as well as the development of explosives and related chemical research and production, led to the Great War later being labelled the 'Chemists' War'. Yet physicists too played a highly significant role. For example, they worked to develop counter-measures to another great innovation of the war, the submarine. Indeed, Rutherford was heavily involved in this project, and turned part of his Manchester laboratory over to the work on underwater sound detection for the Admiralty. Bringing together physicists, engineers, industrialists and military planners working at several sites, this project was a goal-directed enterprise aimed at countering a new German threat to Allied shipping. The outcome was ASDIC (from the Anti-Submarine Detection Investigation Committee, which oversaw its development) – a crude system, initially, but forerunner of the later development of sonar.

The Admiralty was not the only service to benefit from physics in the Great War. The nascent Air Force employed physicists, chemists, engineers and metallurgists at the Royal Aircraft Factory, Farnborough, to explore the properties of materials for aircraft construction and the problems of aerodynamics. The Army employed physicists, mathematicians and engineers to develop ways of detecting enemy gun positions and methods of finding the range and height of enemy aircraft and of controlling guns and searchlights to shoot them down. And all the services experimented with the recent invention that promised to revolutionise battlefield communications and the technique of war: wireless. Physicists from university laboratories up and down the country worked closely with valve manufacturers to develop portable 'wireless telegraphy' sets for use on the battlefield and for communication between ground controllers and aircraft. This in turn made possible the use of aircraft for reconnaissance. Physicists proved their worth, too, in hospitals and field medical units, where X-ray equipment provided valuable diagnostic support in the treatment of casualties. In France, Marie Curie organised a fleet of vans equipped with mobile X-ray equipment – the 'Petites Curies' – to provide radiological support in the field. Not quite on the scale of the Manhattan Project, perhaps, but we can see in these developments the beginnings of the close cooperation between scientists and the military that would later become one of the important ingredients of Big Science.

So: the Great War brought academic scientists, industrialists and the military into close contact for the first time, and it profoundly affected each group. Scientists working in military establishments or units learned to work in collaborative ways on inter-disciplinary goal-directed projects aimed at creating new devices or processes, or at improving existing ones. At higher levels, an extensive committee structure for technical advice and decision-making established senior scientists in policy-making circles. One effect of this was to help establish scientists as key drivers of technical innovation and development in the military services, and to give them a niche as advisers and technical experts. The military became aware of the power of academic science to contribute to the technique of war: as one commentator in 1917 put it, 'whatever university you may choose to visit you will find it to be the scene of delicate and recondite investigations, resulting here in a more deadly explosive, there in a stronger army boot'. At the end of the war, the services therefore consolidated the various research units that had sprung up into a number of research establishments, and during the 1920s and 1930s the Royal Aircraft Establishment, the Admiralty Research Laboratory and the Army's Signals Experimental Establishment amongst others would be important places in the geography of British science.

The Great War also had a significant impact on the scientific relationships between universities, industry and the state. As well as the virtual reinvention of the British chemical industry, the war promoted developments in the newer industries based on electricity and electronics. The General Electric Company established new laboratories at Wembley to work on electronic valves and electric lighting systems, and employed physicists, chemists and metallurgists to explore the properties of materials in efforts to find efficient and durable products. Similarly, companies like Marconi began to specialise in the development of military electronic hardware, and retained close links with university scientists and military research establishments. Though we tend to think of the 'military-industrial complex' as originating after World War Two (more details on this later), in many ways it was already in place during and after the Great War. And if science did not completely determine the outcome of the Great War, it certainly had a profound effect on the thinking of those who would be responsible for future war. It also had a big impact on the organisation of science itself. The Great War saw the creation of new organisational structures and institutions for science, many of which survived the war and helped to shape the development of science in the 1920s. In Britain, for example, the creation of the Department of Scientific and Industrial Research (DSIR) in 1916 led not just to the introduction of a formal national system of science policy and science advice, but to the introduction of grants for research students and awards for more senior researchers – thus creating a career structure for scientists in universities and other research establishments. By 1923, the British government was spending about £1 million a year on physics alone.

One of the most active physics laboratories in the world after the Great War was the Cavendish Laboratory, Cambridge. 'The Cavendish', as it is usually known, has come to have a special place in the history of science as one of the leading centres for nuclear physics in the 1920s and 1930s. Characterised in a 1938 novel by J.B. Priestley as the 'Mecca of good physicists', we often think of it as being the home of the 'pure' physics of atoms, and of a golden age of 'sealing wax and string' – archetypal icons of small-scale, low-budget science. It is true that when Rutherford became head of the Cavendish in 1919 he brought with him the programme of research into the structure of the nucleus established in his last years at Manchester. He turned over a large part of the lab to this work, and organised groups of research students to work on various aspects of his programme, much as he had done at Montreal and Manchester. With his lieutenant James Chadwick, whom he brought with him from Manchester, he quickly made the Cavendish an international centre for nuclear research. As at Manchester, the work was on a table-top scale yet demanded expensive resources and considerable organisational effort. But it is also true that for most of the 1920s, much of this work was mired in controversy, as researchers elsewhere contested the results of delicate nuclear disintegration experiments. Little actual progress was made in nuclear physics at Cambridge in the 1920s.

Indeed, there are other misconceptions about the Cavendish. It is often forgotten, for example, that many of the 20–25 research students a year flocking to the Cavendish from Canada, Australia, New Zealand and elsewhere came to work not on nuclear physics, but on wireless and ionospheric research. A large group working under Edward Appleton (and later Jack Ratcliffe) explored the properties of the ionosphere using wireless technology. Appleton himself was typical of the new breed of physicist who had come to scientific maturity during the Great War. A research student before the war who had been interested in wireless and had become involved in the development of wireless communications for the Army,

Appleton now became a military adviser, sitting on committees overseeing the development of wireless technology for military purposes. He was also sought after as an industrial consultant, and combined both activities with his academic work on wireless and the ionosphere. Later, he would work closely with a DSIR-sponsored research station working on wireless problems for the Meteorological Office and the military. Similarly, Rutherford himself took out patents on his inventions in the field of submarine detection and subsequently became a key government science adviser, serving on numerous committees concerned with the civil, industrial and military uses of science. The story could be repeated many times over. Physics, universities, industry and the military were already closely interlinked in the 1920s.

As soon as our attention is drawn away from nuclear physics, we find that the Cavendish Laboratory in the 1920s and 1930s was interesting for other reasons. In particular, it fostered new ways of *doing* research in physics that entailed not just new buildings but new forms of institutional organisation, new kinds of work and new meanings for scientific experiments. If Dewar and Kamerlingh Onnes pioneered machine physics through their work on low temperatures, the tradition was continued at the Cavendish in the inter-war years by the Russian physicist Peter Kapitza. Kapitza arrived in Britain in 1921, officially as part of a Russian trade delegation. He persuaded Rutherford to take him on as a research student at the Cavendish, and at first worked on problems in nuclear physics. He soon moved on, however, and began to develop new techniques for reaching low temperatures and for attaining very large magnetic fields. This required complex electrical equipment capable of reaching high voltages, and Kapitza exploited the laboratory's links with the Manchester electrical engineering firm Metropolitan-Vickers to obtain what he needed.

By the late 1920s, Kapitza's style of work was beginning to look very different from most of the other researches in progress in the Cavendish. In 1930 Rutherford, whose term of office as President of the Royal Society had just ended, was able to secure £30,000 from the Royal Society to finance the construction of a new laboratory for Kapitza. The 'superlab' was designed by a modernist architect, and its façade included a carving of a crocodile – Kapitza's nickname for Rutherford – by engraver and sculptor Eric Gill. The Prime Minister, Stanley Baldwin, opened the laboratory in 1933 in a public ceremony that was widely covered in the press: a good example of the way in which the managers of the Cavendish used news management techniques to control public perceptions of the laboratory and its work. And it was clear to all that the style of science here was radically different to that elsewhere in British science. A visitor in the early 1930s commented that 'at Professor Kapitza's laboratory, one has to ring at the door to be admitted by a "flunkey" … and you are confronted with Prof. Kapitza seated at a table, like the arch-criminal in a detective story, only having to press a button to do a gigantic experiment'. Big Science indeed.

Kapitza's work at the Cavendish ran alongside more machine building, this time in nuclear physics. By the late 1920s, Rutherford was increasingly concerned about the limitations of using the particles emitted by naturally occurring radioactive substances as projectiles in atomic disintegration experiments. He issued a call for the electrical engineering industry to develop machines that could provide a controllable source of projectile particles. This was the time of spectacular growth of the electricity supply industry in Europe, the USA and the USSR, and the close links between physics and electrical engineering would become invaluable. In answer to Rutherford's call, a former electrical engineer and Metropolitan-Vickers apprentice, John Cockcroft, and his fellow research student Ernest Walton, constructed a series of machines in the late 1920s and early 1930s designed to accelerate sub-atomic particles through high voltages to speeds at which they would be capable of penetrating target nuclei (Figure 7). Again drawing on the expertise and products of the Metropolitan-Vickers company, they succeeded in 1932 in disintegrating light nuclei using artificially accelerated protons. This success – paraded in the media through the efforts of a friendly journalist who acted as unofficial 'press agent' for the laboratory – put accelerators at the forefront of nuclear physics research. Big machines,

*Figure 7.* Machine physics: the particle accelerator with which Cockcroft and Walton artificially split the atom at the Cavendish Laboratory in April 1932.

particle physics, industry and the media began to mesh together (Figure 8).

Elsewhere, too, physicists and engineers were building other kinds of accelerators in what was quickly becoming a race to achieve the highest energies. At Princeton in the United States, Robert van de Graaff built a large electrostatic generator, while at Berkeley, at the opposite end of the country, Ernest Lawrence constructed the first cyclotron, an ingenious machine using electric and magnetic fields to accelerate particles in spiral orbits to high speeds. At

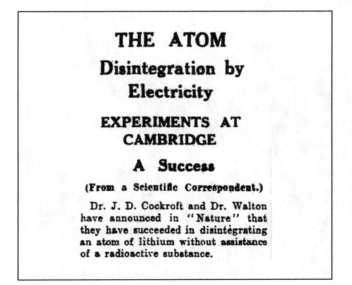

*Figure 8.* A news report of Cockcroft and Walton's work on the 'splitting of the atom', illustrating the early connections between machine physics and media hype.

first, the cyclotron was a table-top piece of apparatus. As the technology proved itself, however, Lawrence and his co-workers built ever-larger versions to reach ever-higher energies. A self-perpetuating cycle set in, establishing a push to larger and larger machines. As one of Lawrence's students put it: 'While we were doing research with the 27-inch and the 39-inch was being designed, he was dreaming up the 60. While we were using the 39-inch and the 60-inch cyclotron was being designed, the 184 was being dreamed of by Ernest Lawrence.' By the mid-1930s, Lawrence and his 'boys' (there were very few women in nuclear physics) were using an 84-ton magnet for their cyclotron – such a huge piece of kit for a physics apparatus that it has been described as the Empire State Building of physics. At the end of the decade, the 60-inch cyclotron had a magnet weighing 220 tons (Figure 9).

In order to achieve these large devices, Lawrence and his team maintained close links with local industry, and as an ambitious scientific entrepreneur, Lawrence was always on the lookout for more resources. In the absence of any significant government funding for American science, 'sponsor relationships' with particular patrons were important; in an area on the boundaries of physics and electrical engineering, where machine-building was at least as important as the physics the machine might produce, patents were also key elements of this new kind of physics. Yet every silver lining has a cloud: it has been suggested that Lawrence's emphasis on engineering

*Figure 9.* Increasingly Big Science: Lawrence and some of his staff posing on part of the 60-inch cyclotron at Berkeley in the late 1930s.

and machine-building led to a lack of rigour in the physics he and his team did with the cyclotron, and that he 'missed' several important discoveries as a result.

After significant changes in experimental technique and the introduction of new kinds of electronic detectors (like Geiger counters), the early 1930s saw a flurry of discoveries in nuclear physics which transformed the prospects for experimental and theoretical research. The discovery of the deuteron (a heavy version of the proton) in late 1931, of the neutron (an electrically neutral nuclear constituent) in 1932 and of the positron (a positively charged electron) in 1933 all gave physicists new tools to explore the construction of the atom and the nucleus. In 1934, the Italian physicist Enrico Fermi and a group of co-workers discovered that elements bombarded with neutrons could be turned into isotopes of other elements – a phenomenon which came to be known as 'artificial radioactivity'. While this phenomenon allowed physicists to explore the properties and inter-relationships of various isotopes, it also had medical implications, because radioactive isotopes could sometimes be used to treat cancer or other diseases. With its controllable beam, the cyclotron became a very good way of producing and exploring the properties of radioisotopes, and also for the creation of therapeutic radiation beams. 'Cyclotroneers' across Europe and the United States now emphasised the potential of their particle accelerators to produce isotopes or other materials for therapeutic purposes, and appealed to medical philanthropies for additional support for their machines. Medicine had its own recent history of development of large machines and special institutes for X-ray and radio-therapy and its own particular funding mechanisms for these large-scale scientific developments, so it was often easier to get funding for medical research than for physics.

By the later 1930s cyclotrons were spreading across the United States and Europe with the support of various universities, industrial firms and philanthropic organisations like the Rockefeller Foundation, which supported American and European physics handsomely in the 1920s and 1930s. Cyclotrons were built in Copenhagen, Paris, Berlin, and in the UK at Liverpool, Birmingham and later Cambridge. It was usually found that Berkeley-trained personnel were needed to help transfer cyclotrons from one place to another (rather like the role of skilled artisans in the spread of textile machinery in the early industrial revolution) – an excellent example of 'tacit knowledge', the need for certain skills in science and engineering to be passed on by showing and doing. Cockcroft-Walton and van de Graaff accelerators, too, began to spread around the growing number of institutions now becoming interested in nuclear physics. Together, these different kinds of accelerators came increasingly to dominate nuclear physics in the 1930s. Trading on links with electricity and electrical engineering companies, and on the skills provided by the booming wireless and electronics industries of the period, the number of laboratories with accelerators multiplied through the decade. As they grew in size and cost, special laboratories had to be built to accommodate the machines that now seemed to drive nuclear physics. Huge funds were required to support these costly new developments – but funding on such a scale was difficult to come by in the Depression years of the early 1930s. Physicists therefore appealed to philanthropists and industry for funding. In 1934, for example, Rutherford and the Cavendish Laboratory successfully persuaded the motor manufacturer Lord

Austin to donate £250,000 towards the construction of a new accelerator laboratory in Cambridge. Like so many others, Cavendish physicists were learning and establishing the realpolitik of machine physics and Big Science.

The new accelerators also brought about changes in social organisation and ways of doing things in physics. They were too large and complicated to be built and operated by one person. Now physicists and engineers began to work closely together in constructing and using the accelerators for 'atom smashing' – the new scientific sport of the late 1930s. As machine physicists increasingly found themselves specialising in one or another aspect of an experiment, teamwork became an important feature, and shifts, rotas and complex forms of division of labour and scientific credit began to replace more traditional ways of working. For younger physicists, accelerators offered exciting new opportunities to be at the cutting edge of technology and physics. For older physicists, the new machines often seemed to be a radical break with their tradition of direct confrontation with nature. Accelerators seemed to substitute specialisation and expense for elegance and economy in experiment: a criticism that was to be made again in the post-war years, as we shall see. For some, the large new accelerator installations even took on the air of science fiction. Commenting on the new accelerator laboratory at Cambridge in the mid-1930s, Rutherford said that the electrical towers rising up to the high roof reminded him of the film of H.G. Wells' *The Shape of Things to Come*.

Clearly, some of the physics laboratories of the 1930s were forerunners of the big post-war particle accelerators – or rather, we might say, they created the conditions that made later accelerators possible. With their high-voltage engineering and close links to the expanding electrical engineering industry, they forged not just a new kind of nuclear physics but also a new model for science generally. There were other examples, too, and not only in university science. In France, for example, the wartime Service Scientifique de la Defence Nationale became the National Office of Scientific and Industrial Research and Invention. In 1922 it merged with the government's science funding body, the Caisse des Recherches Scientifiques, and the two organisations were housed in a 'science city' at Bellevue, near Paris. With lavish government funding, military links, a close relationship between science and engineering, and its large institutional base, this establishment became another home of Big Science in the 1930s. In Germany, the well-funded Kaiser Wilhelm Institutes operated as state institutions bridging academic, industrial and military interests.

Industrial and military laboratories, too, were the breeding grounds of Big Science in the 1930s. With the new importance of aviation and the development of new communications technologies like wireless, close new links began to develop between the various branches of the military and industrial companies. In Britain, military research continued in the 1920s at a relatively low level with research into submarines, aircraft, ordnance and communications technologies. During the early 1930s, however, with a resurgent Germany and the new threat of the long-range bomber, the British government began seriously to consider the air defence of the UK. Military researchers placed large acoustic reflectors around the south coast of England to detect the sounds emitted by approaching bombers, and developed ways of using this early warning to coordinate anti-aircraft defences (both ground-based guns and fighter aircraft). From the mid-1930s, however, this acoustic method was superseded by a new, more accurate and more reliable technique. Government researchers found that radio waves of appropriate frequencies emitted from land stations could be reflected off the surfaces of approaching aircraft (or ships); these echoes could be detected, displayed, and used to coordinate the air defences. In a series of large-scale military exercises in the later 1930s, Radio Direction Finding, later known as *radar*, proved its value as a defensive tool. By 1938, a string of radar stations known as Chain Home was in place along the eastern coast of Britain, with more planned to complete the defensive system. The long-standing connections between academic physicists and electrical engineering companies now came into their own. As companies like Metropolitan-Vickers and Marconi found themselves with large contracts to supply radar equipment to the government, in the summer of 1938 many physicists were inducted into the secrets of radar and took part in exercises designed to test the effectiveness of the Chain Home system. A year later, these people were quickly mobilised and ensured that the defensive radar chain was swiftly put into operation.

We have seen that many of the characteristics of Big Science were in place in the late 1930s. Machine physics, with its massive funding and complex division of labour and work organisation, was becoming widespread. Friendly contacts in the media had been enrolled as allies in the campaign to win funding and scientific and public support for 'atom smashers'. In Britain, at least, physicists were involved in various ways with military establishments, and academic, military and industrial institutions were already working closely together informally, if not formally, on various war-related projects. 'Machines, money, manpower, the media and the military': the five Ms. It was this combination that shaped scientific input into World War Two.

# · CHAPTER 4 ·

# FROM FISSION TO MISSION:
# THE ORIGINS OF THE
# MANHATTAN PROJECT

Nuclear physics developed rapidly in the 1930s. In addition to the construction of ever-larger machines for ripping apart atomic nuclei, scientists used the neutron, discovered by James Chadwick at the Cavendish in 1932, as a tool to explore nuclear structure. By bombarding the nuclei of target elements with energetic neutrons and analysing the products – neighbouring elements in the periodic table – they could deduce the nuclear transformations that had taken place, and therefore work out something of the structures of the various nuclei involved. By the later 1930s, this had become an international game, in which laboratories in Rome, Paris and Berlin competed with each other to wrest the secrets of nuclear transmutation. Personal and national prestige, and perhaps even the possibility of a Nobel Prize, meant that the stakes were high – much like the race to discover the 'W' and 'Z' particles fifty years later. It was out of this series of transmutation experiments that Otto Hahn and Fritz Strassmann found a curious result in December 1938: uranium, when bombarded by neutrons, apparently produced barium, an element half its weight. Hahn called this a 'horrifying conclusion' because it went against all previous experience and knowledge in nuclear physics and chemistry. Puzzled over such an unexpected and unprecedented result, they conferred with their former collaborator Lise Meitner, who had left Germany in 1938 and travelled via the Netherlands to Sweden.

The game of nuclear physics was taking place against the much more serious drama of international political tension. The ascent of the Nazis to power in Germany in 1933 led among other things to the dismissal of Jewish civil servants, including university scientists. In an atmosphere of persecution, many of the Jews (and others whose views were not welcomed by the Nazi regime) fled to Britain, Russia, the United States and elsewhere. Among these émigrés were nuclear physicists like Hans Bethe and the Hungarian-born and German-educated Edward Teller. An Austrian citizen, Lise Meitner had remained in Berlin as long as she could, but the Nazis' annexation of Austria and increasingly overt anti-Semitism finally made her position untenable, and she had sought refuge in Sweden. In Italy, too, Mussolini's race laws drove Enrico Fermi to emigrate to the USA; he left Rome to collect the 1938 Nobel Physics Prize for his work on neutron-induced transmutations and never returned, going on instead to New York where a post awaited him at Columbia University. These exiles and their fear of the Nazis would play an important role in a new race to create, and later control, nuclear weapons.

With her nephew, nuclear physicist Otto Frisch, Meitner reasoned that the bombarded uranium nucleus had split into two roughly equal pieces with the release of a huge amount of energy according to Einstein's famous equation $E=mc^2$ (where E here is the energy released in the splitting of the nucleus, m is the difference in mass between the original uranium nucleus and the sums of the masses of the product nuclei and c is the speed of light). Drawing on a metaphor from cell biology, they named this process *fission*. Following publication of Hahn and Strassmann's work and Meitner and Frisch's speculative interpretation of it, the international community of nuclear physicists rushed to their laboratories to repeat and confirm the counter-intuitive results. By the end of January 1939, fission had been replicated in a dozen or more laboratories in Europe and the United States. Many researchers had also realised one of the implications of the result. If each fission released sufficient additional neutrons, then these neutrons could go on to fission other nuclei in a divergent series, raising the possibility of a self-sustaining chain reaction which could liberate huge quantities of nuclear energy – either in a controlled way, in a reactor, or in a powerful explosion (Figure 10). Experimental work in France and elsewhere soon verified the release of sufficient neutrons to make a chain

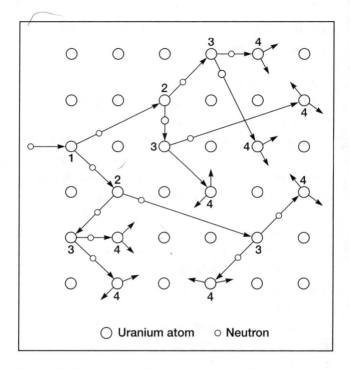

*Figure 10.* The process of nuclear fission and the creation of a nuclear chain reaction. The reaction is initiated on the left (1); each fission here produces two neutrons (2), each of which then produces a further fission (3 ...).

reaction possible. Theoretical work by the respected Danish physicist Niels Bohr and others indicated that of the two principal isotopes of uranium, $^{235}U$ and $^{238}U$ (they differ in the number of neutrons in their nuclei), it was the former that was chiefly responsible for fission. Fission, it now became clear, could possibly be exploited to make a bomb. But most physicists also took the view that the amount of uranium that would be required to extract sufficient $^{235}$U to create a chain reaction and the difficulties of separating the two isotopes of uranium to obtain pure $^{235}$U meant that making such a weapon would, in practice, be nearly impossible.

The gathering storm clouds of war over Europe as Nazi forces mobilised now established a new political context for the fission work. The Germans' invasion of Poland and the declaration of war on 3 September 1939 gave it a new urgency. Though most British physicists had been mobilised to work on the development of radar, there were some who could devote time to the fission problem. Denied access to the secret radar research, émigré physicists Rudolph Peierls and Otto Frisch (Meitner's nephew, whom we have already met) carefully studied the possibility of a chain reaction in uranium. While the mass of natural uranium required to sustain a chain reaction would be measured in tons, they found that the amount of $^{235}$U necessary to create an explosive chain reaction – what came to be called the *critical mass* – was much less than the studies by Bohr and others had suggested. About a kilogram of $^{235}$U, they suggested, would be enough to produce a devastating nuclear explosion. They also worked out a possible technique to separate the uranium isotopes to produce the fissile $^{235}$U in the quantities necessary to create a bomb relatively cheaply and in a reasonable timescale – weeks, rather than the years implicit in Bohr's earlier ideas. Their estimate of the power of a fission bomb was astonishing: a 5-kilogram bomb could have the same effect as several thousand tons of dynamite, and would release enough radiation to be 'fatal to living beings long after the explosion'. Protection against such a devastating weapon would be 'hardly possible'.

Emphasising that Germany might already be developing such a weapon, Peierls and Frisch wrote up their results. Early in 1940, their document found its way to Henry Tizard, a science advisor to the British government. He convened a small committee of leading British nuclear scientists to consider the Frisch-Peierls results and explore the possibilities of a fission weapon. Code-named MAUD, this committee set experimental and theoretical work in train at British universities as evidence began to grow that the Germans were indeed undertaking similar work themselves. Like the scientists setting out to develop radar, the nuclear physicists drew heavily on the connections forged with major industrial concerns like ICI and Metropolitan-Vickers in the 1930s. As their studies of fission progressed, the British scientists working under the neutral-sounding code-name 'Tube Alloys' now began to estimate that an atomic bomb could be produced within three years.

In the United States, the war remained a distant European phenomenon. There, as we have seen, science and government were not closely related. But many of the émigré physicists who had ended up there now worried that the nuclear physicists who had remained in Germany – including the brilliant Nobel Prize-winning theoretician Werner Heisenberg – might work to give Hitler atomic weapons. This fear drove them to make their concerns known at the highest levels of government and to press for secrecy in nuclear work lest helpful information be given to the Germans. Having tried to interest the US Navy in the problem, the Hungarian physicist Leo Szilard persuaded Albert Einstein, himself an émigré now resident in the USA, to write to President Roosevelt urging the government to sponsor investigations into the military uses of fission. An Advisory Committee was established under the directorship of Lyman Briggs (Director of the National Bureau of Standards) and experimental and theoretical work was eventually set in motion on the characteristics of fissile materials and the fission process. At Columbia University, New York, for example, Fermi and numerous collaborators explored the propagation and absorption of neutrons, while related studies of fission were underway at numerous other universities. Fermi began to work on the construction of a nuclear reactor in which he hoped to produce a controlled chain reaction. He used uranium embedded in a three-dimensional lattice of graphite blocks – graphite, studies indicated, would slow down the neutrons and increase the probability of fission. The assemblage took the form of a heap or *pile* – a name which stuck, hence 'nuclear pile'.

In the spring of 1940, the engineer and former

Massachusetts Institute of Technology (MIT) Vice-President Vannevar Bush urged that a new committee be established to promote and oversee defence research in the USA. Headed by Bush and the chemist James B. Conant, President of Harvard, the new National Research Defense Council took over responsibility for the American uranium work. Bush would be a powerful figure in the project from this point on. In September 1940, a group of British scientists headed by Tizard visited the USA. In a quest for US support, they took with them a black box containing many of Britain's wartime scientific secrets, including radar devices and information about the British nuclear programme. From the ensuing exchanges between scientific leaders, it became apparent that the British were some months ahead of the Americans in most areas of fission research. The Americans were ahead, though, in the field of isotope separation. Leaders of the British project now began to think that the United States, with its far larger industrial base and distance from the war, would be a better place than Britain for the development of a nuclear weapon. Official liaison channels were created, allowing extensive contacts between British and American physicists. A month after the German invasion of the Soviet Union in June 1941, the MAUD committee concluded that a uranium bomb was feasible, and issued a report outlining how it could be achieved. Impressed by the results achieved in Britain and by the MAUD committee's optimism, leading American scientists became increasingly committed to the bomb project.

By this time, too, ideas about secrecy had taken hold, and British and American scientists had stopped publishing work on fission in the scientific journals. The sudden absence of nuclear physics in the pages of the *Physical Review* and other periodicals was noticed in Germany – and the Soviet Union – where physicists correctly concluded that American scientists were working in secret to make nuclear weapons. They were making progress, too. In the summer of 1940, researchers at Berkeley had used the cyclotron in neutron bombardment experiments to produce a new chemical element: neptunium (Np, element 93). By February 1941 they had isolated another: plutonium (Pu, element 94). As well as being the first artificially created elements heavier than uranium (hence their designation as *transuranics*), one of them shared an important characteristic with $^{235}$U: plutonium was readily fissile. With this discovery and the confidence imparted by the MAUD report, Bush and Conant lobbied Roosevelt, who gave formal authorisation for full-scale research on the feasibility of an atomic bomb on 9 October 1941.

The entry of the United States into the war after Pearl Harbor on 7 December 1941 only added to the new urgency of the project. By mid-December 1941, the expanded research programme was taking shape, and Bush had assigned responsibility for various aspects of the work, including studies of the chain reaction and weapon theory, methods of isotope separation, properties of fissile materials, and plans for industrial-scale production of the fissile isotopes which would be needed for the weapon. Chain reaction research, widely dispersed at various universities across the United States, was consolidated and moved to the newly constituted 'Metallurgical Laboratory' (an anodyne code name, like 'Tube Alloys') at the University of Chicago. In addition to continuing work at New York and Berkeley, there were also to be dedicated new sites where fissile material for the weapons would be produced on a large scale, and where the bombs would be designed and built. As the project increased in size and scope, its political administrators brought in the military to help coordinate it. The Army Corps of Engineers was placed in charge of the large construction projects that would produce the key materials for the programme and the bomb itself. With its headquarters in Manhattan, New York, the new Corps of Engineers division assumed the innocent code name 'Manhattan Engineer District' (MED) – or, colloquially, the 'Manhattan Project'.

On 17 September 1942, General Leslie Groves assumed command of the Manhattan Engineer District. As deputy chief of construction for the Corps of Engineers, he had been in charge of several large projects, including the construction of the Pentagon in Washington DC. Now he gave new energy and direction to the Manhattan Project. As an engineer with a sound understanding of engineering practice and little time for the order-of-magnitude speculations of the physicists, Groves immediately took a number of key decisions, setting in motion the acquisition of supplies of uranium and the final selection of sites for the industrial production of enriched uranium and plutonium – the raw materials for the weapons. He reviewed all the work in progress and visited the various scientific sites: Chicago's Met Lab, New York's Columbia University, the University of California at Berkeley and others. Groves found a project being pulled in many ways at many different sites, and lacking overall direction. Part of this drift was due to the elaborate committee structure through which the project was organised, and the fact that the scientists were unable to communicate effectively with each other. But part of it, too, was because of an underlying doubt about whether the project was feasible. That doubt was removed by the end of the year.

Early in December 1942, Fermi and his team, now at Chicago, achieved the world's first self-sustaining chain reaction. Planning had begun in May. After substantial difficulties in obtaining supplies of pure uranium oxide and graphite, they had constructed piles of increasing size that allowed them to make systematic measurements of the properties of the fission reaction and gradually work towards criticality. Working in shifts, construction of the final pile, CP-1, began in mid-November under a stand at a university sports ground. On 2 December, the pile, 20 feet high and 25 feet across and containing 771,000 pounds of graphite, 80,590 pounds of uranium oxide and 12,400 pounds of uranium metal, went critical and produced a chain reaction (Figure 11). It had cost $1 million to build. Arthur Compton, head of the Chicago group, telephoned James Conant and in improvised code told him 'the Italian navigator has just landed in the new world'. 'Were the natives friendly?' asked Conant. 'Everyone landed safe and happy.'

Bolstered by the result from Chicago, a December 1942 report from Bush to Roosevelt recommended construction of industrial-scale plutonium (costing $100 million) and $^{235}$U ($160 million) production plants, and heavy water production plants ($20 million). After $85 million had been authorised in June, the total effort was now estimated to cost $400 million. Groves and the Project's scientific leaders selected sites for the industrial operations. An electro-magnetic isotope separation plant for uranium, an experimental pile and a plutonium production plant were to be built at Clinton, Tennessee (the site being generally known as Oak Ridge). The large-scale piles for plutonium manufacture were to be constructed on a large reservation at Hanford, Washington State. After complex and sometimes long-drawn-out land appraisal and acquisition procedures, work on the sites began in the winter of 1942. Construction began at Oak Ridge in February 1943, and at Hanford two months later. Meanwhile, in the spring of 1943,

*Figure 11.* An artist's impression of the first nuclear pile at the University of Chicago, 1942.

Fermi's Chicago pile was dismantled and moved to a new site in the Argonne forest, south of the city, where reactor development studies continued.

At Oak Ridge, responsible for separation of $^{235}$U, a number of modified and much larger versions of Lawrence's cyclotron – known as 'calutrons' – were used. Erected in February 1943, the machines were controlled by 22,000 operators, none of whom knew the real purpose of what they were doing (Figure 12). A gaseous diffusion plant was also established in June, and the Oak Ridge nuclear pile went critical in November 1943. By the spring of 1944, samples of

*Figure 12.* One of the huge calutrons at Oak Ridge used for the separation of uranium isotopes to produce the fissile $^{235}$U.

enriched uranium and plutonium were being produced. The Oak Ridge equipment needed 86,000 tons
of high-quality electrical conductor – which was
supplied in the form of silver on loan from the US
Treasury! The Oak Ridge complex became the fifth
largest city in the state of Tennessee.

At Hanford, development was led by the Army
engineers and the industrial firm Du Pont, who agreed
in November 1942 to build and run a plutonium
production plant and separation facilities under
contract. The Hanford site covered 670 square miles
on the Columbia River close to the Grand Coulee
Dam, an important source of electrical power. The
first pile went critical in September 1944, and
kilogram amounts of plutonium were sent to Los
Alamos early in 1945. For most of the Manhattan
Project workforce, though, the workplace resembled
existing factories – steel mills, chemical works and so
on (Figure 13). Indeed, the Hanford site effectively
became a Du Pont company town, with the same
racial and social segregation that one might find
elsewhere in the United States. In the context of
American wartime labour regulations, many saw the
sites as little more than prisons or forced labour camps
– especially since the unions worked in collaboration
with the bureaucrats. The massive and unanticipated
growth in size and complexity of the project as it
proceeded meant that accommodation for workers
was a constant difficulty. The ensuing problems of
morale led to absenteeism and other difficulties, and

*Figure 13.* The Manhattan Project as factory production: one of the plutonium production piles at Hanford in 1944. There were three such piles along the Columbia River, five miles apart.

meant that worker recruitment and retention were often serious problems for project managers.

In setting up and managing operations at Hanford, Du Pont drew on their pre-war experience of a large technical research and development project: the nylon production programme. The construction of these vast sites was also familiar territory both for the Army Corps of Engineers and for the civilian workers who were drafted in for construction work and operation of the sites. Since the Depression in the United States, gigantic technical projects such as electrification had been seen as both economic boosters and as a key to regional development. One such scheme – the Tennessee Valley Authority (TVA) – was widely held up as an exemplar of what large-scale planned development could achieve. The Manhattan Project's architects had been involved with these schemes in the 1930s and subsequently had strong links with the Army Corps of Engineers. More than that, Oak Ridge was close to the plentiful electricity produced by the TVA, and was in some ways closely modelled on it as an example of a planned, large-scale technical project. Wartime Big Science drew on pre-war large-scale planning. Indeed, it drew so much on the large-scale, almost visionary planning ethos of the TVA that one historian has described the Manhattan Project as 'somewhere between an army base and a utopian social experiment'.

With the massive sites at Hanford and Oak Ridge under development, the choice of the location for the fast fission laboratory remained. At Berkeley, Groves had discussed the fast neutron fission (bomb physics) programme with its coordinator, Robert Oppenheimer. An inspiring theoretical physicist and intellectual, Oppenheimer was trying to hold together work at many places across the USA, an inefficient and time-consuming arrangement. He suggested that all the bomb physics work be concentrated in one isolated laboratory where all the researchers could be in touch with each other, where the programme could be better planned and where security could be better managed.

Despite doubts about his lack of administrative experience, his lack of an authority-conferring Nobel Prize and his left-wing background, Groves appointed Oppenheimer director of the new laboratory (Figure 14). Code-named Project Y, the new laboratory needed to be far inland (away from possible enemy air

*Figure 14.* General Groves, commanding officer of the Manhattan Project, and Robert Oppenheimer, scientific director of the Los Alamos laboratory where the first nuclear weapons were designed.

attacks), far from major population centres (in case of accidents) and topographically isolated (for security). Yet it needed to be relatively accessible, with a local labour force, water supplies and a reasonable climate for year-round work. Oppenheimer suggested a site in the mountains of New Mexico – a ranch school that he had visited during the summers as a child. In November 1942 the Manhattan Project acquired Los Alamos as the site of its nuclear weapon factory.

# CHAPTER 5

# LOS ALAMOS:
# LITTLE SCIENCE ON A BIG SCALE?

While contractors set about developing accommodation, laboratory and workshop facilities on the site,
Oppenheimer began recruiting for Los Alamos – 'the
Hill', as it would come to be called. Partly because of
the remoteness of the site and the likely working
conditions, and partly because many potential
recruits did not relish the prospect of quasi-military
project management, Oppenheimer at first found it
difficult to attract scientists. Groves' agreement that
Los Alamos could operate essentially as a civilian
laboratory eased matters, and Oppenheimer found
that patriotism and a sense of scientific excitement
were enough to persuade many of the best physicists
in the country to join the Project. Among them were
the émigrés Hans Bethe and Edward Teller. Meanwhile, equipment – Cockcroft-Walton and van de
Graaff accelerators, cyclotrons and tons of other
laboratory apparatus – began to arrive from all over
the country. As it gathered momentum, officials
assigned the Project the highest priority rating for

materials procurement – AAA. The codeword 'Silverplate' attached to the Project required instant cooperation from all military personnel – critical in a large wartime economy with many competing demands for staff and resources. The administrative and procurement work of the laboratory was carried out under contract by the University of California, which remained virtually ignorant of the laboratory's real mission for the duration of the war.

In April 1943 Los Alamos opened for business. As the first thirty or so of Oppenheimer's hundred-odd recruits arrived at the shantytown on the Hill from all over the country, they received introductory lectures from Robert Serber inducting them into the Project. Their mission, they were told, was 'to produce a *practical military weapon* in the form of a bomb in which energy is released by a fast neutron chain reaction in one or more of the materials known to show nuclear fission'. Serber told them that one gram of $^{235}$U was equivalent to about 20,000 tons of TNT, estimated the critical mass for $^{239}$Pu as 5 kilograms and that for $^{235}$U as 15 kilograms, and discussed possible designs for the 'gadget', as the weapon was to be called. Many of the new arrivals were startled, because the secrecy of the earlier phases of the project had kept them in the dark about many of the details. Now that they were in on the full secret, they were euphoric. Committees were established and conferences held to plan and oversee the logistics of the laboratory and its research programme. It was agreed that Los Alamos should be expanded to include the work on pluto-nium as well as ordnance development, radically modifying the original conception of a small physics laboratory of about 100 people. For the remainder of the war, the working population of Los Alamos would double in size roughly every nine months; it would become a permanent building site.

The intensely mission-oriented nature of Los Alamos meant that organisation was critical. There were five divisions: Theory (T), Experimental Physics (P), Chemistry (C) [later Chemistry and Metallurgy (CM)], Ordnance and Engineering (E) and Adminis-tration (A). Each division expanded through early 1944, and Groves also brought in more military personnel with scientific or engineering backgrounds to help relieve the labour shortage: by August 1944 they made up 42% of the site's workforce. A rigid hierarchy of communication and responsibility meant that managers and group leaders could maintain strong direction over the work. Strict prioritisation responsive to developments in the research pro-gramme meant that workers could be moved quickly from one part of the project to another as required, and regular meetings assessed the work of the various groups and divisions, reviewed progress, set targets, established priorities and allocated work. Security was a particular bone of contention between scientists and the military. The site was enclosed by two barbed wire fences and was carefully guarded; workers wore colour-coded security badges. Senior scientists were given code names for use within the District. Many of the scientists confronted issues of security – secrecy over documents, limitations on travel and contact with the outside world, censorship and so on – for the first time. One found the conditions so trying that he quit. He wrote to Oppenheimer:

*I found the extreme concern with security morbidly depressing – especially the discussion about censoring mail and telephone calls, the possible militarization and complete isolation of the personnel from the outside world. I know that before long such concerns would make me so depressed as to be of little if any value … I can say that I was so shocked that I could hardly believe my ears when General Groves, director of the Manhattan Project, undertook to reprove us, though he did so with exquisite tact and courtesy, for a discussion which you had concerning an important technical question.*

Despite the fact that scientists' wives were given jobs on the site, recruitment remained a problem over the summer of 1943, not least because of the housing shortage on the Hill. Machinists, shop workers and physicists familiar with engineering were in particular demand. Recruitment of the latter improved late in the year when a new source of supply became available. Fission research had continued in Britain, and the British project leaders learned of the Los Alamos project only in August 1943. There had been concern for some time that the British fission work was vulnerable to enemy bombing, as well as diverting resources from the main war effort. Several of the leaders of the British programme had argued that it should be moved to North America. After high-level discussions between Roosevelt and Churchill that established (albeit loosely) the terms of collaboration, a contingent of British scientists joined the Manhattan Project in December 1943. Though there were only 22 members in the British mission, their experience was to prove vital to the post-war British atomic bomb project. As we shall see, it also helped the Russians in their quest for atomic weapons, for one of the British mission was a German-born émigré physicist: Klaus Fuchs.

Over the summer of 1943, work got underway on the Hill. A major part of the programme was weapon design. From early on, the main idea underpinning thinking on bomb design was that the critical assembly would be produced by firing one sub-critical piece of fissile material into another using a large gun or cannon. This meant a great deal of research on ballistics and the ordnance technique that would allow such a cannon to be fitted into a free-fall bomb. Moreover, the gun mechanism would need to be developed separately for the uranium and plutonium gadgets because of their different characteristics and critical masses. To solve these problems 'ordnance physicists' – physicists with an engineering background or relevant special training – were assigned to this task, under the direction of a naval officer, William S. 'Deke' Parsons, specially recruited to the Hill from radar work. Standard ordnance could not be adapted for the 'gun weapon', so the guns would have to be designed from scratch. Extensive work would be needed on the interior ballistics and on target-projectile-initiator development, and a great deal of experimental testing would be required.

Between the summer of 1943 and January 1944, theoretical work was undertaken on the interior ballistics of various sizes of cannon firing at different speeds and under varied conditions of pressure. The interior ballistics for the expected high-pressure plutonium gun were calculated. A muzzle velocity of 3,000 feet per second would be required, producing a bomb some 17 feet long. Nicknamed 'Thin Man', testing was carried out on the ordnance and fission characteristics of this weapon in the spring of 1944 in an isolated canyon near Los Alamos. Because of its better fission properties, it turned out that the uranium gun weapon would need a much smaller muzzle velocity, and could therefore be shorter and lighter than 'Thin Man'. 'Little Boy', the uranium weapon, would be 6 feet long and weigh less than 10,000 pounds – an easy load for a Lancaster or a B-29 bomber.

As work moved towards the finalisation of the gun designs, research continued on other aspects of the weapons. A key part of the devices would be the *initiator* – a device using the rare and difficult-to-handle radioactive element polonium to produce an initial flood of neutrons to help kick-start the chain reaction at the moment of assembly of the critical mass. Much work was needed on initiator development, especially in producing sufficient amounts of polonium of adequate purity. Again, industrial resources could be drawn in. In mid-1943, Oppenheimer had recruited Charles Thomas, director of research of the Monsanto Chemical Company, to coordinate the plutonium work and improve communication between the various labs (for example, by holding inter-laboratory meetings and helping to resolve the problems of plutonium allocation between labs for the many tests that had to be carried out). Now Oppenheimer also placed polonium production in the hands of Monsanto. In a typical piece of Manhattan Project organisation, Monsanto in fact took major responsibility for the polonium work, creating a new lab for the purpose. The initiator design was settled by the summer of 1944.

The accelerators that had dominated late 1930s nuclear physics laboratories now came into their own, because they could be used to create particles or isotopes, whose properties could then be explored. Several accelerators were moved to Los Alamos from various university laboratories, and work soon underway included experiments by groups under Robert Wilson on the cyclotron, John Williams on the van de Graaff machine, and John Manley on a Cockcroft-Walton accelerator. The skills of instrument-builders and traditional small-scale scientists were vital, too. Darol Froman led a team developing electronics equipment, Emilio Segrè led a group working on the properties of radioactive and fissile materials, Donald Kerst directed work on a small test nuclear reactor, Hans Staub and Bruno Rossi developed particle detectors. Indeed, detectors and electronics were central elements of the overall research programme, because it was through them that the critical measurements could be made. So important were they that researchers were poached from high-priority radar work to help make fast, reliable and effective devices. This group grew rapidly from ten workers in early 1944 to 80 a year later.

The head of P Division, Robert Bacher, planned an ambitious programme with multiple approaches ('redundancy') to obtain as much information in as short a time as possible. One unfortunate consequence was that the experimental programme was beset by inconsistent and conflicting results, for example over the values for the average number of neutrons emitted by uranium and plutonium in fission. Experimentalists learned too of the problem of 'capture' of neutrons – their absorption by other nuclei at certain 'resonance' energies. Part of the difficulty was that experimentalists were still trying to work with tiny samples of material from diverse sources, and they keenly awaited larger samples of fissile material from the production sites. At Chicago, chemists under Glenn Seaborg were working to purify the minute samples of the new element plutonium and to develop ways of manipulating and characterising tiny quantities of material. In the summer of 1943 one of the first physics experiments at Los Alamos measured the number of neutrons emitted by a 200-milligram sample – a barely visible speck – of Seaborg's plutonium when it fissioned.

It was also because of the shortage of fissile materials that theoretical analyses of the behaviour of neutrons in fast fission reactions and experimental measurements on very small samples of material to determine their characteristics would be extremely important. The theoretical physicists of T Division therefore played a prominent role, both in predicting likely outcomes and in interpreting results in the light of theoretical frameworks. They made extensive use of mechanical calculating machines to solve numerical problems, estimate the effect of varying parameters, extrapolate tables of results and so on. They tried to use flexible models that could readily be changed in the light of each new piece of experimental data. Key problems included understanding the ways in which neutrons diffused in the fission process over time (which required the development of numerical approximation methods); calculation of the critical masses for the gun devices; and estimation of the likely effects of the 'gadgets'. Here, too, multiple approaches allowed them to check results from one theoretical method against those of another, and so improve confidence in the final outcome. There was also constant checking against experimental data and revision of calculations in order to make sure that theory and reality were synchronised. Even for the theoreticians, making bombs was intensive work.

Much of the actual work of bomb making revolved around the production and manipulation of the fissile elements: messy, dangerous work. The chemists and metallurgists of CM Division developed new ways of making, improving and handling materials (for example, new uranium casting procedures), means of purification and characterisation of plutonium, and ways of heat-treating steel for use in the gun weapon. Because only microgram amounts of plutonium were at first available, CM Division kicked off with a big effort on uranium, devising procedures for producing, purifying and handling metallic $^{235}U$ (whilst avoiding criticality), exploring the properties of uranium compounds for possible weapon use and of uranium solutions for use in reactors, determining sample densities (very important for critical mass calculations) and so on. Some of these studies would turn out to be dead ends, but this was part and parcel of the redundancy approach. The plutonium work was the most challenging, for the element was still new, had not yet been isolated in bulk form, and its properties were largely unknown. It was therefore widely dispersed across the different Manhattan Project sites. One effect of this was to create a certain amount of competition between them (for example, to be the first to produce plutonium metal).

By the end of 1943, work was forging ahead on all fronts, driven by Groves' determination. At Los Alamos, studies of various kinds were in progress, with optimism about the basic bomb design tempered only by continuing shortages of fissile material. Though there were certain fixed parameters to the gadget – it had to fit into the bomb bay of a B-29 bomber, the largest that would be available to the US Air Force in 1945 – much remained unknown about the basic properties and behaviour of fissile materials, chain reactions and other internal aspects of the gadget. In particular, measurements of the critical constants of plutonium (for example, its density, crucial for calculating its critical mass) were ambiguous and often contradictory. But the industrial production plants at Oak Ridge and Hanford were on schedule to produce gram amounts of enriched uranium and plutonium in 1944, which would allow the missing information to be obtained.

However, just as all seemed to be going smoothly, with the uranium and plutonium weapon designs well advanced, an unexpected problem emerged which threatened half of the laboratory's work. In the spring of 1944, the Los Alamos programme faced a major crisis that would put it and the entire Manhattan Project organisation to the test.

## · CHAPTER 6 ·

# THIN MAN BECOMES FAT MAN: THE PLUTONIUM IMPLOSION PROGRAMME

When gram quantities of plutonium from the experimental pile at Oak Ridge started arriving in the spring of 1944, the Met Lab and Los Alamos chemists were able to achieve a better understanding of its properties. It turned out to be much more complicated than uranium, which they had been using as a surrogate in the absence of concrete information. It had several unexpected characteristics, including a much lower melting point than expected and the fact that it could exist in different crystalline forms. But Emilio Segrè's group also discovered that the first samples of $^{239}$Pu from the nuclear piles at Oak Ridge displayed a capacity to fission spontaneously five times higher than the samples of cyclotron-produced $^{239}$Pu that had been used in previous experimental tests. Extended studies over the following weeks confirmed the result. They quickly worked out that the neutron flux in the nuclear pile had created $^{239}$Pu mixed with a hitherto unobserved isotope of plutonium, $^{240}$Pu. $^{240}$Pu had different characteristics, including a higher spontaneous fission rate and a neutron emission rate five times higher than that of $^{239}$Pu. To the bomb-builders this was alarming because the weapon might now predetonate – i.e. 'fizzle' rather than explode. At a stroke, the 'gun' method therefore became unviable for $^{239}$Pu, and an alternative technique would be necessary.

Studies of spontaneous fission had begun under Segrè at Berkeley in late 1941, using delicate and sophisticated electronic counting equipment. Several researchers had thought of the possibility that $^{239}$Pu might capture a neutron in a reactor to form $^{240}$Pu, but no one had suspected that its spontaneous fission rate would be so much greater than that of $^{239}$Pu. In June 1943, the group had moved to Los Alamos and continued its work as part of Oppenheimer's 'cover all angles' philosophy. The work was difficult because of the small samples of $^{239}$Pu available and the delicacy of the measurements – the group worked in a remote canyon some distance from the main site to keep its electronic equipment free from vibration and background radiation. As increasing amounts of plutonium arrived from the Berkeley cyclotron, they were able to make better measurements, though still not as accurate as they needed. Now, in April 1944, the bad news about plutonium and spontaneous fission spread rapidly. Met Lab scientists were at first sceptical, not least because the result meant that much of their work on plutonium separation and characterisation would have been in vain. At Los

Alamos, the realisation quickly set in that the gun method could not now be used for plutonium, since the assembly of the critical mass would be too slow to avoid predetonation. The use of a faster gun and the possibility of separating the unwanted $^{240}$Pu from the $^{239}$Pu were ruled out as impracticable within the Project's timescale. Yet the Hanford piles were about to produce substantial quantities of plutonium for weapon use. After much agonising, the Project's managers decided in July 1944 to halt work on the plutonium gun and purification programme immediately. Instead, they reoriented the plutonium work around another concept for the bomb design: one of the lab's sidelines – an insurance-policy line of research on a technique known as 'implosion'.

The idea of implosion had been suggested by Seth Neddermeyer at the beginning of the Project in April 1943. It involved using a blanket of explosives to crush a hollow shell of fissile material in on itself, creating a critical mass and hence an explosion. This arrangement would allow faster assembly of the critical mass than the gun design, and would therefore perhaps help avoid the danger of predetonation. A small implosion group was established under Neddermeyer. Drawing on the expertise of the military Explosives Research Laboratory, it spent much time on a mesa a few miles from the main site carrying out test explosions on steel tubes, trying to create symmetrical implosions of the kind that would be necessary in a weapon. But at that stage the implosion programme was given low priority, as the gun design was expected to work for both uranium and for plutonium.

In September 1943 the Hungarian émigré mathematician John von Neumann visited Los Alamos. With long experience of work on shaped explosive charges, he suggested changes to the implosion experimental programme. His idea of using a higher ratio of explosive to metal mass (giving faster assembly) appealed to Oppenheimer and others, because it suggested the possibility of using a smaller amount of less pure fissile material than would have been needed for the gun device. As we have seen, fissile materials were scarce, so any way of using them economically and efficiently was welcome. Neumann's interest and prestige raised the priority of the implosion work, but so too did his specific and detailed plan for diagnostic experimental and theoretical work. Implosion was given higher priority in November 1943. George Kistiakowsky, head of the Explosives Research Laboratory, was brought in as consultant to the expanded experimental effort. Its key task was the measurement of parameters such as the symmetry, time of collapse and degree of compression of test implosions. This involved the use of high explosives as precision tools, and meant that the implosion group now developed much stronger links with industry – already characteristic of much research elsewhere on the Hill.

Kistiakowsky arrived to work on the implosion programme in January 1944. In June he replaced

Neddermeyer as head of implosion experimentation. Oppenheimer tellingly felt that Neddermeyer's style was too 'academic' to run the programme, which now involved hundreds of workers: the original implosion group had 'not yet become accustomed to the idea of large scale operation'. Oppenheimer and Kistiakowsky now initiated a major reorganisation of the implosion work, including more planned and systematic experiments with better diagnostics, and a strong push on the development of instrumentation to make the required measurements under the difficult conditions of the tests. Much of this development was interdisciplinary, bringing together chemists, metallurgists, theorists, machinists, electronic engineers and explosives experts. Logistics, careful planning and strong management: all were crucial to the reorganisation of the implosion programme, and all those involved would remember the lessons.

As part of the shift in priorities, theorists also now switched to spend more time on implosion. An implosion theory group was set up in January 1944 under Teller (the group included the physicist Richard Feynman). Drawing on suggestions by Rudolf Peierls, a member of the British mission working in New York, the group made extensive use of punch-card calculating machinery specially ordered from IBM to integrate blast wave equations and to compute the critical masses of bodies with various geometries. Such calculations could suggest important information about the characteristics of implosion, but despite the energy and dedication of their female operators the mechanical calculators were limited in speed and volume: it was partly these limitations that led to the development of fast electronic digital computers.

The more the scientists explored it, the more implosion seemed to have serious problems. The tests were plagued by 'jets' (tongues of molten material which disrupted the symmetry of an explosion) and 'spallation' (the breaking off of bits of material under the pressure detonation waves). The group drew on more British advice from James Tuck, who suggested using a three-dimensional explosive lens (Tuck was an assistant to Lord Cherwell, Churchill's science adviser, and drew on work being done on related problems in the UK). The US physicist Luis Alvarez also suggested the use of high-quality electric detonators to obtain simultaneous detonation at several points. The problems of producing explosive lenses and detonators would be central to the implosion work right through until the summer of 1945. More experience was brought in. Norris Bradbury was recruited from a Navy Ordnance post to help with lens research, and there was extensive and systematic testing of various detonators supplied by US industry to find the most appropriate and reliable. Outside expertise continued to be important. Another British physicist, Geoffrey Taylor, questioned Los Alamos thinking about the interface between core and tamper in the implosion assembly, leading to a complete redesign of the implosion gadget in the summer of 1944.

By this time, of course, Segrè's news about spontaneous fission had sunk in. There was a drastic and far-reaching reorganisation of the laboratory in late July 1944 to give implosion top priority. This was helped by the fact that much of the nuclear physics programme and work on the uranium gun weapon were far enough advanced to allow staff to be moved around. Laboratory activities and responsibilities were now organised in a completely different way, moving some groups into new divisions, and splitting some groups up entirely and reassigning them in different combinations. In the new divisional structure, a Gadget (G) Division was responsible for experiments on the critical assembly of active materials, hydrodynamics of implosion, and the design and construction of the tamper and core. The Explosives (X) Division now took responsibility for design of the explosive components of the implosion bomb, and for methods of detonating high explosives, including casting, lenses and explosive charges. Ordnance (O) Division had responsibility for the final weapon design and delivery, and also for coordination of laboratory activities with the military groups carrying out flight tests at Wendover, Utah, and other operational matters. Research (R) Division absorbed much of former P-Division, especially the groups working on the production and measurement of neutrons. T-Division was little affected by the reorganisation, apart from the bolstering of the implosion work. Bethe continued to insist that T-Division staff work closely with those in other groups so as to be in intimate contact with experimental work and results. CM-Division's workload was considerably lightened by the discontinuation of plutonium purification work, and some staff were reassigned. Partly because Fermi joined Los Alamos from Chicago in September 1944 and his experience and abilities qualified him for a Divisional leadership role, a new Fermi (F) Division brought together miscellaneous projects. Hundreds of people and a complex programme of work reorganised – all because of the discovery of a small physical effect in plutonium.

While the implosion programme got into gear, the uranium gun programme was now consolidated and streamlined in Group O-1, which could focus all its energies on systematic testing and on contingency planning, as well as turning the design into a deliverable weapon. The design was a conservative one, rigorously thought-out and tested. In August 1944 the first large shipment of $^{235}$U arrived at Los Alamos from Oak Ridge. Production was not proving easy, raising worries that not enough of the 'product' would be available for weapon use in the war. Production picked up by January 1945, however, and tensions eased. The Los Alamos metallurgists now began to focus on producing and working metallic $^{235}$U. Work continued on initiator development, in conjunction with Monsanto and other industrial collaborators. At the same time, small-scale tests of gun design were undertaken, including drop tests of various models of the uranium bomb.

By the end of 1944, the crash implosion programme was well underway. It commanded a multi-disciplinary effort of fourteen or more groups, whose studies encompassed the theoretical mechanics of implosion and the use of iterative experiments to explore the use of high explosive lenses as precision tools. But the overall state of implosion research in autumn 1944 was discouraging. Because of problems with the implosion weapon, and anxious to ensure success at all costs, General Groves demanded a higher priority for the gun programme. He set a target completion date of 1 July 1945 – the expected date when enough uranium and plutonium would be available for the weapons. Among the physicists there was optimism that this target could be met. But because of the many uncertainties now involved in the implosion programme, a test was now thought to be important for the implosion gadget, even though it would waste precious plutonium. By the autumn of 1944 a test site had been selected in New Mexico, in the region called Jornada del Muerto – Journey of Death. The test of the plutonium device would be called 'Trinity'.

## · CHAPTER 7 ·

# FROM TRINITY TO VICTORY: MAKING AND USING THE FIRST NUCLEAR WEAPONS

As the Manhattan Project entered 1945, there was a shift from thinking of the uranium gadget Little Boy just as a piece of mechanism to thinking of it as a deliverable bomb. Now began the process of planning for the problems of delivering and preparing Little Boy for combat use. The design of the gun mechanism was settled, and the main task now was to find the best ballistic shape for the bomb casing and a reliable and accurate fuse that would detonate the bomb at a fixed altitude. Again, there was much systematic testing of various designs and planning for contingencies in the event of failure of any of the many components. After intensive work, Little Boy was ready for combat by early May 1945. Now it only awaited enough $^{235}$U and completion of its initiator. In June, the fissile material arrived, and was turned into bomb components using the procedures carefully developed over the preceding months. Its makers were so confident in it that they did not think a test was necessary. At the end of July 1945, Little Boy was ready for delivery.

By this time, the implosion programme, too, had essentially achieved its goal. Following the reorganisation of the laboratory, intensive research had been undertaken into the characteristics of plutonium and implosion, again drawing on advanced diagnostic techniques to obtain the necessary data. The physicists still needed information about the dynamics of implosion and about the different kinds of explosives and lens systems. Seven diagnostic methods were employed in parallel to study implosion, as well as detonators, lenses and initiator. Again, instrument development was key, and often relied on developments elsewhere – for example, fast-sweep oscilloscopes from the radar programme that allowed measurements to be made on a timescale of tenths of microseconds. At the same time, empirical 'try-it-and-see' methods were the basis of much of the experimental work, especially on the high explosives needed for the lenses. The British contingent were involved in various aspects of the lens work – Ernest Titterton on fast detonators, Klaus Fuchs on the analysis of jets and spalls, and Rudolf Peierls on theory of explosive lenses. Achieving a completely symmetric implosion remained the key problem. The theory group responsible for numerical calculations played an important role in making calculations of likely actual implosion dynamics.

Meanwhile the chemists were still working on the properties of plutonium and ways of handling the larger quantities expected imminently from Hanford.

Because of the change of design, the purification standards had now been relaxed, and metallurgists worked out a large-scale process for producing plutonium metal. Much of this work involved finding suitable materials in which to handle and fabricate this new element. Again the 'cut-and-try' style of experimentation and systematic trials were crucial in the investigation of the various factors important in determining the behaviour and characteristics of plutonium and its compounds and alloys. Like other aspects of the programme, outside expertise was imported when necessary. The implosion researchers nevertheless felt themselves to be working against the clock in the development of production methods. The first large shipment – more than 100 grams – of Oak Ridge plutonium arrived in January 1945, followed by increasing quantities of Hanford material. Again, communication within the Manhattan Project was far from perfect, and there was initially some friction between Los Alamos and Hanford over the quality of material and secrecy. These difficulties were soon ironed out, however, and the Hanford plutonium was deemed up to specification by Los Alamos chemists and metallurgists.

From early 1945, much of the implosion work was dictated by design questions and the increasingly pressing need to fix design for bomb production. One particularly significant question arose from a suggestion by theoretician Robert Christy that a solid sphere rather than a hollow shell of plutonium be used in the core of the implosion device. The 'Christy gadget', while less efficient than the hollow shell device, was a more conservative design. Though contentious, it was approved by Oppenheimer and Groves, and soon became the agreed design basis for the implosion weapon. Again, a detailed plan of attack was established, with work allocated to the various groups and deadlines set. As in the gun programme, the emphasis now shifted towards development and testing of the actual weapon. A new division was created to take responsibility for preparations for the planned test of the implosion weapon, scheduled for the summer of 1945. The increased amounts of fissile materials on the site now also meant that health physics and decontamination instrumentation assumed a new prominence.

Through the spring of 1945, frantic work continued on development of the implosion weapon. The initiator work was given high priority, though the continuing lack of polonium meant that the choice of initiator design relied heavily on theoretical work. Again, opinions differed about the best design, and some of the physicists worried about this choice right up to the Trinity test. Much of the work on explosive lenses was still of a 'trial and error' nature as the workers scaled up from models to full-scale components and encountered unexpected problems like cracking after casting. Facing their own scale-up problems and shortages of staff, the metallurgists worked hard to produce hemispheres of plutonium for the test device and for the combat weapon that would presumably follow. Relying on scheduled deliveries of plutonium from Hanford brought problems of its own as deadlines persistently slipped; in the event, deadlines for delivery of components for both the test device and the bomb were met. While the experimental physicists checked and re-checked results on critical masses and neutron behaviour, multiple researches continued in parallel on detonators, outside contractors Raytheon helping with firing circuits. Much of the work here involved testing various designs for robustness and reliability. As with other elements of the implosion programme, work – and doubt – continued almost until the test itself.

Plans for the test of the implosion device had been in train for many months. By early July 1945, they were approaching their climax and all the logistics were in place. As the design programmes had come to an end, many of the scientists now focussed their attention on the test. Hundreds of measuring instruments, specially designed cameras and other equipment were placed at various positions and distances around the designated test site to record as many aspects of the anticipated explosion as possible. Diagnostic instrumentation would allow the scientists to assess the performance of the device itself, and especially those elements such as the detonators over which they had had most difficulty and doubt. As in most other aspects of the Project, a review committee was established to decide which experiments should be performed and who should be responsible for them. Several ways were approved of measuring one of the most important outcomes of the test: the destructive power of the device. Theoreticians – including British physicist William Penney, later to head the UK atomic bomb programme – had used data from bomb damage in Britain to estimate the likely blast effects of the test and the combat weapons, including probable radiation effects. In May 1945, a rehearsal test took place with 100 tons of ordinary explosives and a small amount of radioactive material to allow the scientists to calibrate their instruments, check the organisation of the test and the planned measurements and uncover any weaknesses in their plans. Now, in yet another instance of the Project's characteristic forward-looking and self-protective bureaucracy, seismic measurements of the effect of the explosion, though not essential, were approved in case of legal action for blast damage by residents of nearby towns.

By 13 July the implosion device had been assembled at Los Alamos and it left for the test site. The plutonium core (or 'pit') travelled separately and was inserted at the test site that afternoon. In the evening the assembly was hoisted up the test tower. Concerns about unfavourable meteorological conditions were overridden by political considerations: the new President, Harry Truman, was due to meet with Stalin and Churchill at Potsdam on 16 July, and wanted the test results before the meeting. On the night of 13

July, there were thunderstorms at the test site, which could create difficulties for the many experiments planned around the test. Early the following morning, however, the storms cleared and the winds were favourable to prevent any fallout problems. At 05.30 on 16 July, the device was exploded. The sky lit up with a dazzling flash of white light. A cloud began to rise from the purple-red fireball of the explosion, assuming the form first of a raspberry then of a goblet or mushroom (Figure 15). The sky glowed orange in the new dawn of the nuclear age.

Despite later accounts which often emphasised the observers' philosophical or moral reflections on the Trinity test, the scientists' immediate reactions were first of euphoria and second of trying to obtain as much information as possible about the explosion

*Figure 15.* The result of the Manhattan Project and the beginning of the nuclear age: the first nuclear explosion at the Trinity test site, 16 July 1945.

from their carefully planned experiments. Some of the diagnostic experiments did not work because of a much higher than expected yield of radiation from the explosion: but the very fact that the bomb had worked indicated to the jubilant scientists that its mechanics were satisfactory. Other measurements over the next days and weeks allowed them to calculate the yield of the device at about 20,000 tons of TNT (three times larger than predicted), with a damage radius of about 1,000 yards. The largest bomb dropped in the war to that date had been the British 'Grand Slam', which delivered 12,000 pounds. The experiments on the yield and blast at Trinity allowed the group in charge of combat delivery to fix parameters for the operational use of the weapons. The crash implosion programme had paid off. The 'moment of truth' at Trinity 'demonstrated that a well-funded, large-scale, mission-oriented, multi-disciplinary research laboratory employing the new blend of pre-war and Manhattan Project strategies could handle a problem that only one year earlier looked impossible'. More significantly, so did Hiroshima and Nagasaki.

The Manhattan Project had originally been established to counter the threat of Nazi nuclear weapons. Following the allied invasion of Europe in the summer of 1944 and the end of the European war in June 1945, the Project administrators decided that the bombs, due to be completed in the summer of 1945, could still have a role in the continuing war against Japan. After

VE Day a few scientists left, but the Manhattan Project continued its work. As early as July 1944, plans were being laid for a Pacific forward base for a nuclear offensive against Japan, and by August B-29 bomber crews were being trained to drop atomic bombs. In February 1945, the Pacific island of Tinian was chosen as the base for nuclear-capable B-29s; by June, bomb components were being shipped from Los Alamos to Tinian and training missions were taking place using dummy weapons. During late July, scientists and the bomb cores arrived on Tinian and assembly of the combat weapons began.

A target selection committee had been established in the spring of 1945, and had chosen two primary targets and subsidiary backup targets in Japan, carefully exempting them from the US Air Force's carpet-bombing campaign of Japanese cities. There had been stiff opposition to the decision to drop the bombs on cities from the Chicago scientists, who had found themselves at something of a loose end after the wind-down of the plutonium work and had turned to thinking about broader moral and political questions surrounding atomic weapons. The Chicago group had argued that:

> A demonstration of the new weapon might best be made, before the eyes of representatives of all the United Nations, on the desert or a barren island. The best possible atmosphere for achievement of an international agreement could be achieved if

> *America could say to the world, 'You see what sort of a weapon we had but did not use. We are ready to renounce its use in the future if other nations join us in this renunciation and agree to the establishment of an efficient international control.'*

But with the thought of having to justify the huge expenditure on the Project to Congress, the emerging prospect of the Soviet Union as a post-war adversary, and perhaps even a wish to test the effect of the weapons on cities (there was strong anti-Japanese sentiment in the USA in the wake of atrocities in the war in the Pacific), Groves was adamant that the bombs should be used in combat.

In the early hours of 6 August 1945, the B-29 Enola Gay, piloted by Colonel Paul Tibbetts, took off from Tinian. At 08.15 local time, it dropped the uranium gun-type weapon Little Boy over the Shima Hospital in the city of Hiroshima. The bomb exploded with a yield of approximately 15,000 tons of TNT (Figure 16). Of the city's population of about 280,000 civilians and 43,000 military personnel, 78,000 were killed outright by blast and fire, with a further 37,000 subsequently missing. As radiation effects set in, by the end of 1945, the deaths totalled 140,000; by 1950, 200,000. When the Enola Gay returned to Tinian, Tibbetts was immediately awarded a Distinguished Service Cross. A White House press release later that day described the bombing as 'the greatest achievement of organised science in history'. Secretary of War Henry Stimson

*Figure 16.* The impact of the Manhattan Project: the effect of 'Little Boy', the uranium bomb, on Hiroshima. 220,000 were killed. The United States authorities ensured that many of the photographs released did not feature victims and survivors.

called it 'the greatest achievement of the combined efforts of science, industry, labor and the military in all history'. At Los Alamos, the scientists celebrated.

Three days later, a second B-29, Bock's Car, piloted by Major Charles Sweeney, left Tinian carrying Fat Man, the plutonium implosion weapon. Finding its primary target, the arsenal at Kokura, obscured by haze and smoke, it flew on to its secondary target: the industrial port of Nagasaki. It too was obscured by cloud, and the crew attempted a radar approach. At the last minute a gap opened up in the cloud, however, and they were able to make a visual approach to a stadium a few miles from the original aiming point. At 11.02 local time on 9 August, Fat Man exploded over Nagasaki with a blast later estimated at 22,000 tons of TNT. By the end of 1945, 70,000 people had died; by 1950, 140,000. As Manhattan Project planners and scientists began to make preparations for sending more plutonium cores to Tinian for further attacks, the Japanese began surrender negotiations the day after the bombing of Nagasaki. They surrendered on 15 August 1945.

At Los Alamos, there were celebrations and parties at the successful outcome of over two years' work. Yet pride mixed with guilt as the results of the bombings became known in detail. As scientists began to consider their post-war career options, the technical work of the laboratory slowed and almost ground to a halt. Oppenheimer, like many others, chose to leave the laboratory to return to the academic world. Some chose to stay on: Norris Bradbury, Oppenheimer's successor as director, argued that the laboratory should continue its work to give America a strong international bargaining position. More tests would be crucial to improve weapon design and to provide intellectual stimulus to weaponeers. According to Bradbury, another Trinity test 'might even be FUN'.

## · CHAPTER 8 ·

# AFTER THE BOMB: BIG SCIENCE AND NATIONAL SECURITY

In the shadow of Hiroshima and Nagasaki, nuclear science gained an awesome reputation. The devastating impact of the bomb had important consequences for national security and for the scientists who had made it. The Manhattan Project had ended up costing $2.2 billion, and had become as large as the US automobile industry. During the war, the physicists and other scientists who had worked at Los Alamos and the Project's other sites had learned a great deal about effective ways of doing research. Their science had been pragmatic and goal-directed, with an emphasis on speed and results. The techniques they had become familiar with involved trial and error, scale models and iterative procedures. They had learned to work in multidisciplinary teams involving experimentalists and theoreticians, scientists and engineers. These teams had worked on individual projects that were carefully managed: group leaders needed managerial skills to coordinate groups of interdisciplinary staff. They had experienced scientific decisions about priorities and allocation of responsibilities being taken by committees reflecting both scientific and military aspects of the larger project. Funding had not been a constraint, given the Manhattan Project's 'Silverplate' priority rating, so that problems could often be tackled in several ways at once. The scientists now used these new experiences and skills as they returned to their laboratories after the war.

Of course, the Manhattan Project was not the only wartime research and development project to use scientists and engineers on a massive scale – though its outcome was undoubtedly the most dramatic and the most deadly. The various radar projects in the UK and the USA developed similar approaches with multidisciplinary teams, and also saw the development of new connections between science, industry and the emerging national security state. Organisations comparable to Los Alamos had emerged, bringing together scientists and engineers to work on the complex electronics required for navigation and the detection of enemy aircraft and shipping: in the UK, the Telecommunications Research Establishment at Malvern, and in the US the MIT Radiation Laboratory, which had 4,000 staff by 1945. By the end of the war over \$3 billion had been spent on radar research and equipment. All these projects gave the scientists who participated in them an insight into new and effective ways of organising research. Hierarchical groups, targets, organisation charts, teamwork, security, personal and group empire building and links to military, political and financial authority now became a standard part of the managerial physicist's practice. In America particularly, a new kind of researcher emerged, bringing together the skills of the physicist, equipment-building engineer and entrepreneur and the ability to mobilise and manage human, financial and technical resources. This researcher was characterised by a pragmatic, utilitarian approach bringing together theory, experiment and engineering with a stress on getting results. Many of them also emerged from the war with a string of new industrial, military and governmental contacts that they would put to good use in their post-war work.

For scientific administrators and politicians, the Manhattan Project had demonstrated the value of scientists and of large, goal-directed projects. At the end of the war, according to one scientist:

*Suddenly physicists were exhibited as lions at Washington tea parties, were invited to conventions of social scientists, where their opinions on society were respectfully listened to by lifelong experts in the field, attended conventions of religious orders and discoursed on theology, were asked to endorse plans for world government, and to give simplified lectures on the nucleus to Congressional committees.*

If physicists were the aristocracy of post-war science, the Manhattan Project seemed to have shown the value of investment in basic research on the fundamental properties of matter. Who in the mid-1930s would have thought that pure nuclear physics would have such devastating application as that unleashed on Japan in 1945? During the war, US federal funding for science had increased from $50 million to $500 million per year. The increase continued after the war. New institutions were created to manage this expenditure: in the US the Office of Naval Research (ONR), the National Science Foundation (NSF) and the Atomic Energy Commission (AEC) were important channels of federal funding. By 1949, 60% of funding for physics in American universities was supplied by the Defense Department. Total levels of spending were at $1 billion by 1950, and at $3 billion by 1956. But more significant even than the amounts involved was the fundamental shift to government funding of science.

The links forged between physical scientists and the military during the war, and particularly in the Manhattan Project and the radar projects, were to prove crucial to the development of post-war science. Early in 1946, General Eisenhower, then the US Army Chief of Staff, circulated a memo on 'Scientific and Technological Resources as Military Assets'. Eisenhower argued that the military now needed 'to support broad research programs in educational institutions'. The US Navy had reached a similar conclusion. It established the ONR in 1945 to forge links between the Navy and university scientists. The

ONR pursued its mission aggressively, and became a significant sponsor of university research in the USA. Already by 1946, for example, the ONR had distributed $24 million in contracts, involving 602 academic research projects and 4,000 scientists and graduate students. Universities became adept at orienting themselves towards the possibilities of military funding, particularly in aeronautical and nuclear engineering, underwater acoustics, optics, space science and electronics. And university scientists liked the ONR because ONR contracts respected academic styles of research and allowed their projects to be unclassified and publishable.

Federal patronage for science at universities often took the form of contract research centres whose objectives and research plans were determined by the military sponsors but which were managed by the university as contractor. Examples included the Applied Physics Laboratory at Johns Hopkins University in Baltimore and the Jet Propulsion Laboratory at Caltech in Pasadena, supported first by the US Army and later by the National Aeronautics and Space Administration (NASA). These centres usually worked on military problems and weapons technologies, and operated under military-style access and classification procedures, often putting them at odds with the rest of the host institution. Elsewhere, large units not formally contracted to the military operated within universities – for example, the Cornell Aeronautical Laboratories in Buffalo, or Charles Stark Draper's

Instrumentation Laboratory at MIT. At MIT, the Radiation Laboratory of Electronics (RLE) continued the tradition of the wartime Radiation Laboratory in studying microwave radiation, supported by a $600,000 contract from the ONR, the US Air Force and the Signal Corps. At Stanford University, ONR similarly financed work on microwave technology. The Stanford Electronics Research Laboratory was established in 1947, and by 1950 Stanford was carrying out nearly $500,000 of electronics research for the Department of Defense. Again, the Korean War had a major impact. After an offer of further funding by the Navy, the Applied Electronics Laboratory was established, and it was soon carrying out $1 million of research. At Stanford, 'the department of aeronautical engineering ... became a virtual adjunct of Lockheed and its missile program'.

The military was clearly a pervasive influence in the post-war physical sciences in the USA. By 1952 James Conant, still a powerful figure in science administration, noted their 'almost fanatical enthusiasm' for research and development. 'The Defense Department, in regard to research,' he observed, 'is not unlike the man who sprang on to his horse and rode off madly in all directions.' Much of the US government's support for science after the war went to research in non-nuclear areas. In terms of organisation and the use of resources, much of this funding supported small science, but on a very extensive scale. This fundamentally shaped the development of basic research.

At Stanford, for example, the development of microwave electronics for the Navy 'also contributed to the linear electron accelerator with which Robert Hofstadter earned a Nobel Prize by exploring the interior of the atom ... the Nobel Prize-winning work of Felix Bloch on nuclear magnetic resonance ... [and] ... the equipment that made possible UHF television broadcasting'. In the UK, too, many of those who had led wartime projects found themselves well placed to exploit wartime experiences and contacts to obtain funding. For example, Bernard Lovell at Manchester University was able to draw on his connections with the wartime radar project to obtain equipment and skilled staff for the development of radio astronomy. This led to the construction of the large steerable radio telescope at Jodrell Bank in Cheshire, one of the icons of Big Science in post-war Britain.

So: the post-war years saw what has been called the 'mutual embrace' of scientists and the military. It would be wrong simply to see the military as exploiting science in these years: scientists and scientific administrators actively sought out military funding for their projects – and won it, often on their own terms. Nor was the relationship between science, the military and the state always happy: Robert Oppenheimer had his security clearance withdrawn following a series of hearings in 1954 over allegations that he had been linked to Communists. Yet for the most part physicists, in many ways the new high priests of technoscientific culture, struck an implicit bargain with the state, in which they accepted funding for their work but were expected to be ready to turn their advice, their skills and their large, costly equipment to the ends of war. This expectation is well exemplified in the field to which we now turn our attention once again: accelerators and high-energy physics.

## · CHAPTER 9 ·

# FROM BIG SCIENCE TO MEGASCIENCE: THE AGE OF THE ACCELERATORS

In the USA, the Manhattan Engineer District was succeeded in 1947 by the Atomic Energy Commission, a civilian body with responsibility for all nuclear matters. The AEC inherited the various Manhattan Project sites, and turned them into 'national laboratories' where applied nuclear research continued, built on the Manhattan Project model. These laboratories at Los Alamos, Berkeley, Oak Ridge and Argonne carried on the wartime work on bombs and reactors, and followed through many of the questions that could not be addressed during the war when the goal of producing nuclear weapons was so pressing. They became permanent features of the American scientific landscape. The AEC competed with the military agencies to fund nuclear research: 40% even of early ONR funding went to support research in nuclear physics. In this context, nuclear physics could only win.

As in other fields, scientists who had experienced both the largesse of federal support and the goal-directed nature of scientific development during the war now became adept at playing the system to gain funding for their projects. The exemplar: Ernest Lawrence. At Berkeley, Lawrence's laboratory was already a large institution with big machines and private funding. Indeed, as we have seen, in many ways it was the archetype of Big Science, providing a model for the organisation and style of research in the Manhattan Project. Soon after the war, Lawrence persuaded Groves to commit Manhattan Project funds for the development of 'pure' nuclear research at Berkeley. Emphasising the argument that discoveries in pure physics before the war had made the atomic bomb possible, Lawrence and others now used the relationship between fundamental research and possible military applications to justify large expenditures on accelerators and other projects in high-energy physics. Where Lawrence's pre-war laboratory had a few buildings, about 60 staff and a budget of up to $125,000 – large for its time, as we have seen – he now began to think in terms of a budget of millions of dollars and a staff of hundreds. This was a measure of the shift produced in him and in science by the Manhattan Project. It also set the pace for the continuing development of Big Science.

The mid-1940s saw the development of new kinds of accelerator, and concentrated work on them now became possible. For example, the betatron accelerated electrons in circular orbits (unlike the cyclotron, which accelerated particles in spiral orbits). The synchrotron used small, moveable magnets synchronised with particles' orbits – its size was therefore not limited by the size of the magnet, like the cyclotron, but only by the available money and land. The cyclotron could also be modified in the same kind of way to produce what was called a synchrocyclotron. This modification was developed for the 184-inch cyclotron at Berkeley soon after the end of the war using Manhattan Project funds. Lawrence's first 4-inch cyclotron in the USA in 1930 had an energy of 80,000 eV.* By 1939, Lawrence's 60-inch machine was reaching 19 MeV. Using the new principle the 184-inch Berkeley synchrocyclotron reached 195 MeV in late 1946 – double the energy that had been planned for it in 1940. The AEC also supported the development of other kinds of accelerators and a nuclear chemistry laboratory at Berkeley – all to be run by Manhattan Project veterans.

By this time, accelerators had become powerful symbols of institutional prestige and vehicles for scientific competition. The developments at Berkeley after the war prompted physicists on the US east coast

------

*The energies of these machines are measured in electron Volts, usually abbreviated to eV. One electron Volt is the energy gained by one electron as it passes across a potential difference of 1 Volt. The much higher energies of particle accelerators are expressed in megaelectron Volts (millions of eV, abbreviated as MeV), gigaelectron Volts (billions of eV, abbreviated GeV) or teraelectron Volts (trillions of eV, abbreviated TeV).

led by Isidore Rabi to request their own large-scale particle accelerator facility: this became the Brookhaven National Laboratory, established in 1947 under a consortium known as Associated Universities, Inc. on a former US Army base on Long Island, New York. Brookhaven's prime piece of equipment was to be an AEC-funded synchrotron delivering more than 1GeV. It would also support work on nuclear reactors for the AEC's national reactor programme – both for weapons material production and for exploration of power and medical uses of reactors. At Berkeley's near neighbour, Stanford University, physicists and engineers drew on their wartime electronics expertise to build a 33 MeV linear accelerator (usually abbreviated to linac): this reached 75 MeV by 1950, and 900 MeV by 1957.

Within the United States in the late 1940s, then, there was significant competition between universities for federal funding for large accelerator projects, and between federal agencies to fund them. One wag commented on the relationship between physicists and the military in a song called 'Take Away Your Billion Dollars':

> *Up on the lawns of Washington the physicists assemble*
> *From all the land are men at hand, their wisdom to*
>   *exchange.*
> *A great man stands to speak, and with applause the*
>   *rafters tremble.*
> *'My friends,' says he, 'you all can see that physics now*
>   *must change.*

> *'Now in my lab we had our plans, but these we'll now*
> *expand,*
> *Research right now is useless, we have come to*
> *understand.*
> *We now propose constructing at an ancient Army base,*
> *The best electro-nuclear machine in any place.*
>
> *'Oh – it will cost a billion dollars, ten billion volts*
> *'twill give,*
> *It will take five thousand scholars seven years to make*
> *it live.*
> *All the generals approve it, all the money's now at hand,*
> *And to help advance our program, teaching students*
> *now we've banned.'*

Despite such satirical comment, the scientific context for machine physics was changing in significant ways in the late 1940s. Cosmic ray research was continuing along the pre-war model, using cloud chambers and photographic plates. Though the work continued to be organised in small groups, the limited number of opportunities for experimentation (e.g. the small number of suitable mountain sites) meant that there was increasing large-scale international collaboration. For example, a 1953 project based on Sardinia involved researchers from 22 different labs in twelve countries. Out of all this work, more and more data was being produced, so that analysing the thousands of photographs became a much more complex task – which developed into what was a 'cottage industry' by the later 1940s. This intensive but widely distributed work was rewarded by the discovery of new particles in 1947 – the pi-meson (or pion) in Bristol, and the so-called 'V-particles' at Manchester. This increase in the number of particles naturally had implications for the particle accelerators.

By the late 1940s, the new generation of particle accelerators was beginning to reach high enough energies to produce the particles observed in cosmic rays. For example, the new Brookhaven synchrotron – designated the 'Cosmotron' – began operation in 1952; by early 1953 its energy of 3.3 GeV allowed physicists to make pions and V-particles and to explore their properties and interactions (Figure 17).

*Figure 17*. Heralding the age of the large accelerators: the cosmotron at the AEC's Brookhaven Laboratory in the early 1950s.

By this time, there were also synchrocyclotrons at Harwell and Liverpool in the UK, Amsterdam, Uppsala and in the USSR. There were electron synchrotrons at Glasgow, Oxford and in the USSR. And there were proton synchrotrons at Brookhaven and Birmingham, with bigger machines under construction at Berkeley and Dubna, near Moscow. The competition between laboratories characteristic of pre-war nuclear physics now became enmeshed with national security issues, as the major scientific states all began to equate nuclear physics with weapons and power. Indeed, accelerator physics in America and Europe in the mid-1950s was given a major boost by the inauguration of the Dubna accelerator in 1956, promoting fears of an 'energy gap' comparable to the 'bomber gap' and 'missile gap' of the raging Cold War. The race to higher energies had intensified.

As the Cold War became entrenched, work on nuclear weapons continued in earnest in the various US national laboratories. In the UK, too, the Atomic Energy Research Establishment (AERE) at Harwell was part of the UK's plan to develop nuclear physics and nuclear energy. Many of the British scientists who had worked on atomic weapons and radar went there. Among Harwell's scientific projects were a 110-inch, 175 MeV cyclotron commissioned in December 1949 and nuclear research reactors. In 1947, denied access to US atomic information (despite their wartime collaboration) the UK government decided to establish its own nuclear weapons programme. The Atomic

Weapons Research Establishment (AWRE) was set up at Aldermaston to coordinate this work. The British nuclear programme relied heavily on the scientists who had been involved in the Manhattan Project, such as William Penney (who led the UK programme) and theoretical physicist Klaus Fuchs. Because of his extensive knowledge of the work at Los Alamos, Fuchs made important contributions to the British project, and was instrumental in its success in 1952. But Fuchs was also a Communist, and passed information from Los Alamos to the USSR. Aided by Fuchs' treason, the Soviets' testing of an atomic bomb in 1949 took the West by surprise, and led to still further investment in military research and nuclear science (as well as to security crackdowns in the military-scientific establishment).

After the outbreak of the Korean War in 1950, government sponsors of university research found themselves under great pressure to justify basic research. They called on the patriotic obligations of award holders and redirected basic research programmes towards military ends. At MIT's Research Laboratory of Electronics, for example, a new project and a large new laboratory – Project Charles and the Lincoln Laboratory – were established with a huge budget to consider the feasibility of a strategic early-warning radar system. At Berkeley, Lawrence turned parts of his lab over to more directly military-oriented projects, particularly work on the 'Super' or hydrogen bomb. This device used the enormous energy from a fission reaction to fuse together light nuclei, with an even greater release of energy. Work had been carried out on the idea at Los Alamos during the war, though by February 1944 it was realised that it would be a much more difficult project than originally thought. Edward Teller nevertheless relentlessly promoted the 'Super', and in June 1944 he was given his own remit to work on it. The 'Super' programme was marginal to the creation of the first atomic bombs, but it provided important information on deuterium reactions and energy transfer in fission reactions. This was valuable to work on the hydrogen bomb in the early 1950s. In re-mobilising his laboratory, Lawrence created a new Materials Testing Accelerator at Livermore to gener-ate neutrons, needed for the production of fissile materials for weapons. As the cost of this scheme escalated to $21 million in 1952, however, Livermore was turned over to work on the hydrogen bomb. By 1954 Livermore had more than 1,000 staff. The United States tested a thermonuclear device in 1952, the USSR following in 1953 and the UK in 1957. With the advent of the H-bomb, hundreds of times larger and more destructive than the nuclear weapons that had been dropped on Japan, the world entered the thermonuclear age.

The intensification of military research during and after the Korean War dramatically increased the importance of applied research in universities. But as government and military support for Little and Big Science began to dominate, some scientists began to question the effect of military support on academic science. Leading science administrators worried that university science was becoming too programmatic, that it was being 'warped' by the military, and that universities were becoming like industrial corporations. These tensions in the mid-1950s were resolved by the Soviets' launch of the satellite Sputnik in 1957. In the subsequent fears of a 'missile gap', Western governments' support for science soared again, with particular emphasis on support for science education and space research. Scientists – especially nuclear physicists – continued to trade on the prestige which their subject had gained during the war and which was sustained by their involvement in work on the hydrogen bomb in the early 1950s. In this context, the growth of machine physics was an extension of what had been happening before the war, but now at a greatly accelerated rate.

It was in the context of the Cold War too that discussions took place within a United Nations and UNESCO framework in the early 1950s about a collaborative European physics laboratory. Though this has often been seen simply as a response to the growing cost of accelerators, there are other important factors. With fears of Communist influence in European governments, Marshall Aid and other American initiatives sought to bind Europe into an Atlantic alliance. The European movement led to the creation in 1948 of the Organization for European Economic Cooperation, which led in turn to the Council of

Europe and the European Coal and Steel Community – forerunners of today's European Community. NATO – the North Atlantic Treaty Organisation – was created in 1949 as a vehicle for integrating the defences of the USA and Western Europe. It was in this political and ideological context that US physicist Isidore Rabi proposed the creation of a European laboratory for particle physics. Rabi had Brookhaven in mind as a model for a large, collaborative institution. After two years of technical and diplomatic discussions, CERN – Conseil Européen pour la Recherche Nucléaire, or European Organisation for Nuclear Research – was established in 1952. After much discussion a site near Geneva was chosen and the laboratory was established in 1954 by twelve founding member states: the Federal Republic of Germany, Belgium, Denmark, France, Greece, Italy, Norway, The Netherlands, the UK, Sweden, Switzerland and Yugoslavia (Yugoslavia left in 1961; Austria joined in 1959; Spain joined in 1961, left in 1969 and rejoined in 1983; Portugal joined in 1985).

CERN's founders planned to build a scaled-up version of the Brookhaven Cosmotron, working at an energy of 10–20 GeV, with an estimated cost of $210,000. When a CERN team visited Brookhaven in August 1952, however, they learned of a very recent development in accelerator design: a new arrangement of the magnets, in a system called 'alternating gradient' or 'strong focussing', allowed much more cost-effective use of materials and space. The principle was adopted for the planned CERN synchrotron, which it was now expected could achieve 20-30 GeV. While design and development work was underway, CERN commissioned a 600 MeV synchrocyclotron in 1957, and used it to observe the decay of pions into electrons and neutrinos. The strong focussing proton synchrotron began operation at 25 GeV in 1959. To some extent, though, the accelerator development work was at the expense of actual physics research. Relatively little thought had been given to the kinds of experiments that might be carried out on the new machine. Thus the Brookhaven synchrotron retained the lead in physics terms. Close behind it was a new 6 GeV proton accelerator – the Bevatron – at Berkeley. Physicists there (including Manhattan Project veteran Emilio Segrè) used this machine in 1954–5 to discover the antiproton in an explicitly mission-oriented project that in some ways mirrored the crash programme at Los Alamos in 1944–5.

The highest-energy accelerator laboratories now became like the laboratories undertaking radioactivity research at the beginning of the century, with researchers travelling to a central location ('facility') to use an immobile resource. And as the machines became ever larger and the researchers more numerous, the institutions housing them became organisationally more complex. A typical laboratory would now contain overlapping groups of physicists and engineers who designed and built the accelerators; physicists who did experiments with them on various kinds of detectors and produced scientific data; and physicists who theorised about the meaning and significance of those results. There would also be a large support staff responsible for maintenance and operation of the accelerator itself, including people responsible for delivering particle beams on demand; people for ensuring computer systems were operational; and all sorts of other services associated with large institutions – as we saw earlier with the organisation of Fermilab.

The increasing complexity of accelerators in the 1950s was matched by the increasing sophistication and complexity of their subsidiary instrumentation, especially the detectors the scientists used to observe the particles as they interacted with each other. A new kind of detector – the bubble chamber – and new techniques for processing large numbers of photographs created the conditions for new discoveries. Major shifts in material and organisational culture accompanied the introduction of new techniques. The new detectors were capable of taking thousands of photographs of particle interactions, and these had to be carefully scanned and interpreted for indications of novel phenomena – or 'events', as they would come to be known. Teams of 'scanners' – often women, as with the Manhattan Project calculators – used complex equipment at widely distributed locations to carry out this work. If the cosmic ray research of the 1940s had been a cottage industry, high-energy physics of the 1950s and after was mass production.

By the late 1950s, not least because of the massive increase in scope of detectors, a host of strange new particles had made their appearance. To account for the new 'zoo' of particles, Murray Gell-Mann and other theoreticians devised a new theory of matter and a new set of fundamental units of matter with even stranger properties: quarks. These quarks themselves became the focus of enquiry for the next generation of machines in the 1960s. By this time, too, further developments were taking place in accelerator design. Physicists conceived the idea that, instead of accelerating a beam of particles and firing them into a stationary target, two beams could be accelerated in opposite directions, and made to collide with each other. The resultant collision would take place with twice the energy of the two beams (two moving cars colliding with each other at speed do more damage than one car crashing into a wall), and would produce new phenomena for study. The 'collider' suggestion raised significant technical problems to keep the engineers and design physicists busy, but also raised the possibility of very interesting new physics.

While physicists at Stanford developed the Stanford Linear Accelerator Center (SLAC), commissioned in 1965 with an 18 GeV electron beam, their colleagues at CERN exploited the collider concept with the Intersecting Storage Rings (ISR) plan – two interlinked 25 GeV proton accelerators that could act as a particle collider. After much technical and political discussion – always the bane of such large international schemes – the project began in 1965, and was commissioned in 1971. Meanwhile in the USA, physicists responded by beginning to lay plans for a new national accelerator laboratory at Batavia, near Chicago. With a $250 million construction budget, the new accelerator was planned as a 6.4 km ring operating at 200 GeV. Symbolically, its main building was modelled on a thirteenth-century French cathedral. The new laboratory was founded in 1969, and its accelerator commissioned in 1972, first operating at 200 GeV and later that year at 400 GeV. In 1974 the laboratory was renamed the Fermi National Accelerator Laboratory – usually known as Fermilab. Already, though, federal funding for high-energy physics – the US was spending $143 million a year on it by 1963 – was tied to institutional and regional politics, as well as national prestige. In the USA, high-energy facilities became part of the 'pork-barrel' politics by which federal funds and projects are distributed to particular regions for political reasons. As we shall see, this was to have important consequences.

In what was now becoming a transatlantic tit-for-tat competition, CERN staff started laying plans for its next-generation machine. In 1971 approval was given for a 7-km collider ring: the Super Proton Synchrotron (SPS), initially planned for 300 GeV. While this machine was being planned, CERN boosted the energy of its existing proton synchrotron. When the

SPS was commissioned in 1976, it exceeded its design specifications: by the end of 1978, its peak energy had been taken to 500 GeV. But the huge cost of the SPS – 1,150 million francs – had serious implications for the participating countries' own national accelerator programmes in the face of limited science budgets. In the UK, for example, continued participation in CERN meant the sacrifice of the country's own high-energy machines at the Daresbury and Rutherford Laboratories, so that its high-energy physicists now became completely reliant on CERN. Apparently unaffected by such considerations, US physicists at Fermilab now began to make plans for the biggest machine yet – the Tevatron, using superconducting magnets and aiming at the phenomenal energy of 1 TeV. (1 TeV is about the energy of motion of a flying mosquito, but in an accelerator such energies are packed into a space about a million million times smaller than a mosquito.)

All these developments took place against the background of significant changes in the theory of high-energy physics. In 1967 Steven Weinberg and Abdus Salam put forward the now-accepted model of the fundamental structure of matter – the 'Standard Model'. This theory predicted the existence of two heavy particles, which came to be known as the W and Z particles (the W was expected to exist in two forms, W+ and W–). The search for these particles was to drive accelerator development in the 1970s. By this time, it was clear that the machine builders were in a competitive and self-perpetuating spiral to higher and higher energies in the quest for particles – and Nobel Prizes. More broadly, however, Big Science was flourishing elsewhere too. The Manhattan Project had been the prototype for the creation of NASA and the US manned space programme, and the success of the Apollo missions in the late 1960s was a public triumph for large-scale mission-oriented research. Though the development of manned space flight was undoubtedly carried out by the aerospace industry in conjunction with more military-oriented projects (for example the construction and operation of spy satellites), NASA became a major supporter of university research from 1963, with its 'Sustaining University Program' distributing $221 million to 175 institutions from 1963 to 1971. Through this and other programmes, space research, microwaves, lasers and computers became key elements of post-war science and engineering, both in their own right and as tools for other sciences.

# · CHAPTER 10 ·

# THE INVENTION OF 'BIG SCIENCE': LARGE-SCALE SCIENCE AS PATHOLOGICAL SCIENCE

Not everyone was happy with the development of Big Science and its implications for scientific policy and scientific practice. In the 1940s, Merle Tuve, a prominent American physicist, had abandoned nuclear physics because he felt that it had 'changed from a sport into a business'. In 1959 he published an article entitled 'Is Science too Big for the Scientists?' in which he suggested that people were coming to serve the machines, rather than the other way round. In 1950s Britain, too, the astronomer Fred Hoyle argued that the UK should not develop its own space programme because the intellectual value of space research was not worth the money and labour which would be needed to carry it out, and would certainly not justify the diversion of resources from other areas of science. It was in this context that Alvin Weinberg, Director of the AEC Oak Ridge National Laboratories, used the term 'Big Science' in 1961 to refer to what he saw as the pathological development of science.

In a paper on the 'Impact of Large-scale Science on the United States', Weinberg argued that though they represented the 'supreme outward expressions of our culture's aspirations', the institutions of Big Science had raised difficult practical and philosophical problems. Likening the projects of Big Science to the 'monumental enterprises' through which earlier civilisations had expressed their aspirations – the pyramids of the Ancient Egyptians, the cathedrals of the Middle Ages (remember the design of Fermilab?), Versailles in seventeenth-century France – Weinberg famously suggested that:

*When history looks back at the twentieth century, she will see science and technology as its theme; she will find in the monuments of Big Science – the huge rockets, the high-energy accelerators, the high-flux research reactors – symbols of our time just as surely as she finds in Notre Dame a symbol of the Middle Ages. She might even see analogies between our motivations for building these tools of giant science and the motivations of the church builders and the pyramid builders. We build our monuments in the name of scientific truth, they built theirs in the name of religious truth; we use our Big Science to add to our country's prestige, they used their churches for their cities' prestige; we build to placate what ex-President Eisenhower suggested could become a dominant scientific caste, they built to please the priests of Isis and Osiris.*

But Weinberg also suggested that the economic distortion caused by the construction of such monuments in many cases contributed to the decline of past civilisations. He asked three important questions about Big Science: (1) 'Is Big Science ruining science?' (2) 'Is Big Science ruining us financially?' (3) 'Should we divert a larger part of our effort towards scientific issues which bear more directly on human well-being than do such Big Science spectaculars as manned space travel and high-energy physics?' His answers to each are revealing of the dissent among scientists and policy-makers about the way in which science was going. Let's consider each in turn.

First, as to whether Big Science was ruining science, Weinberg argued that Big Science needed publicity to justify the public's support. This led to the development of a distinctly journalistic tendency in Big Science, with important scientific and technical issues tending to be argued in the popular, rather than the scientific, press. The line between journalism and science was becoming blurred, he said, and 'the spectacular rather than the perceptive becomes the scientific standard'. Thus for Weinberg, the use of the media by scientific leaders to promote Big Science and their own laboratories' interests was a pathological feature of science (as we have seen, however, this practice was already taking place in the 1930s if not earlier). He also criticised the enormous growth of administration in science. But there was an even deeper criticism. In one of the 'most insidious effects of large-scale support of science', Weinberg saw 'evidence of scientists spending money instead of thought'. With money plentiful, he argued, 'there is a natural rush to spend dollars rather than thought – to order a $10^7$ nuclear reactor instead of devising a crucial experiment with the reactors at hand, or to make additional large-scale computations instead of reducing the problem to tractable dimensions by perceptive physical approximation.' 'The line between spending money and spending thought,' he concluded, 'is blurring.' More amusingly, the science commentator Daniel Greenberg created the mythical academic figure of Dr Grant Swinger of the Center for the Absorption of Federal Funds (motto: 'something always comes along') to capture this strange situation!

In answer to his second question, 'Is Big Science ruining us financially?' Weinberg answered 'Yes.' He pointed out that annual US federal expenditure on science was currently (1961) running at $8.4 billion: about 10% of the entire federal budget. Over 80% of this sum was for military research and development. This budget had increased by about 10% per year over the previous decade. At that rate, he argued, 'we shall be spending *all* of our money on science and tech- nology in about 65 years. Evidently something will have to be done or Big Science will ruin us financially.' So what was that something to be? Weinberg suggested that a small percentage of the federal budget – less than 1% – be allocated to non-military science over a period of about fifteen years. The science budget would then increase in line with the gross national product, and scientists would have to make choices about which sciences to support:

*These choices, which will require weighing space against biology, atomic energy against ocean-ography, will be the very hardest of all to make – if for no other reason than that no man knows enough to make such comparative judgements on scientific grounds. The incentive for creating a favorable public opinion for a pet scientific project will become much greater than it now is; the dangers of creating a political 'in' group of scientists who keep worthy outsiders from the till will be severe. Nevertheless, it is obvious that we shall have to devote much more attention than we now do to making choices between scientific projects in very different fields.*

This led Weinberg to his third question, 'Can we divert the course of Big Science?' Conceding that in the context of the Cold War it would be 'naïve, if not hopeless, to argue that we should not use scientific achievement as a means of competing with the USSR', Weinberg nevertheless argued that neither of the two iconic projects of Big Science – space and high-energy physics – were suitable events for these 'scientific Olympic Games'. Space exploration was hazardous, costly and not relevant to the lives of most people, despite popular interest in manned space travel.

Similarly, high-energy physics, though perhaps more scientifically interesting, was remote from human affairs (though it had the advantage that in this field the USA was ahead of the USSR). No, Weinberg's own battleground of choice would be 'scientific issues which have more bearing on the world that is part of man's everyday environment, and more bearing on man's welfare, than have either high-energy physics or manned space travel'. One such area might be molecular biology, and Weinberg thought that 'the probability of our synthesizing living material from non-living before the end of the century is of the same order of magnitude as the probability of our making a successful manned round trip to the planets'. He also suspected that 'most Americans would prefer to belong to the society which gave the world a cure for cancer than to the society which put the first astronaut on Mars'.

Such choices, Weinberg concluded, had become 'matters of high national policy':

*We cannot allow our overall science strategy, when it involves such large sums, to be settled by default, or to be pre-empted by the group with the most skilful publicity department. We should have extensive debate on these over-all questions of scientific choice; we should make a choice, explain it, and then have the courage to stick to a course arrived at rationally. ... [W]e must not allow ourselves, by short-sighted seeking after fragile monuments of Big Science, to be*

*diverted from our real purpose, which is the enriching
and broadening of human life.*

Coming from the director of a large national laboratory, itself one of the citadels of Big Science,
Weinberg's critique was enormously influential. He
followed up his article in *Science* with more talks and
papers, which were subsequently collected in a book
of essays, *Reflections on Big Science*. Again he com-
plained about the 'corruption' of science by Big
Science. 'In this age of big money for science,' he
argued, 'it is harder than in earlier days to find the
scientist dedicated solely to truth; he is responsible for
spending Big Money, and his pursuit of science is
sometimes distorted by his method of funding.' Visiting MIT in 1962, for example, Weinberg remarked
that it was hard to tell 'whether the Massachusetts
Institute of Technology is a university with many
government research laboratories appended to it or a
cluster of government research laboratories with a
very good educational institution attached to it'.
In the same year, retiring President Dwight D.
Eisenhower in his farewell address drew attention to
the changes in military organisation that had taken
place during the first decade of the Cold War. He
expressed his concern at the emergence of an
'immense military establishment' working hand-in-
glove with a massive arms industry – what he called
the 'military-industrial complex'. His concern was
with the 'unwarranted influence' that industrialists

could bring to bear on national economic and strategic policy, and with the political and moral implications of the 'technological revolution of recent decades'. As he put it:

*In this revolution research has become central; it also becomes more formalized, complex, and costly ... Today, the solitary inventor, tinkering in his shop, has been overshadowed by the task forces of scientists and laboratories and testing fields. In the same fashion the free university, historically the fountainhead of free and scientific discovery, has experienced a revolution in the conduct of research. Partly because of the huge costs involved a government contract becomes virtually a substitute for intellectual curiosity. For every old blackboard there are now hundreds of new electronic computers.*

Of course, Eisenhower's own policies, especially after the shock of Sputnik in 1957, had helped bring this situation about. But he captured, too, something of the autonomous dynamic of Big Science, the feeling that 'the machine' was beginning to run away with itself. He warned that while the 'prospect of domination of the nation's scholars by federal employment, project allocations, and the power of money is ever present and is gravely to be regarded', citizens should also 'be alert to the equal and opposite danger that public policy could itself become the captive of a scientific-technological elite'.

Such criticisms were highly relevant to big physics. By the mid-1960s, there was anxiety about the increasing sums being spent on accelerators in what one observer called the 'cargo cult of modern science'. When Fermilab was being discussed in 1967, the *New York Times* pointed out the widespread objections to 'the expensive irrelevance of a 200 billion electron volt accelerator to any real present national problem', such as the Vietnam War or social tension in American cities. There was similar criticism in Europe. In the late 1960s, as European physicists pushed for the 300 GeV SPS at CERN, British scientists and politicians debated the impact of the $400 million tab on the UK's own high-energy physics programme, leading to the closure of two UK national accelerators, as we have seen. One British politician wrote in his weekly column for the magazine *New Scientist* that two things had sunk into the consciousness of Members of Parliament about the high-energy physicists: 'that they are working in a field of central importance to the advancement of human knowledge; and that they are unloved among their scientific colleagues on account of their voracious appetite per head for resources. (I shudder to think how many biologists could be financed for the cost of a proton synchrotron or a synchrocyclotron.)'.

Within the scientific community – even within the physics community – there were tensions over the ever-increasing cost of HEP. In 1971, the prominent condensed-matter physicist Philip Anderson (who would share the 1977 Nobel Physics Prize for his work on the structure of magnetic and disordered systems) asked, 'Are Big Machines Necessary?' Anderson pointed out that no results from high-energy physics had been relevant to the rest of science for a generation. He criticised high-energy physics for allowing itself 'to be sold as a field relevant to nuclear power – as is clear, for example, from the fact that the AEC ran the accelerator laboratories'. He disputed the high-energy physicists' claims that their science was 'fundamental' to all other sciences, arguing that in aggregate form matter obeyed different laws than it did as fundamental particles, so that progress even in other branches of physics was in no way dependent on that in high-energy physics. Many of the most important results in recent high-energy physics, he suggested, could have been obtained at much less expense, and often without big machines at all. More pointedly, 'the money and brains which go into such an object as the Batavia [Fermilab] accelerator are very likely to be wasted from the point of view of the very purpose for which they were intended. These are not the first expensive instruments which don't mean very much but they are now doubly dangerous in that other valid lines of research are being closed down'. 'To do an experiment just because it is feasible is all right,' he concluded, 'but not if you are effectively closing down tens of hundreds of equally valid investigations.'

Big Science as a *label*, then, was born of the worries in the 1960s that large-scale science, especially high-energy physics, was getting out of control. It came as a prelude to broader attacks on physics and the academic-military-industrial complex in the wake of the Vietnam War and a decline in public confidence in physics, which to many was now associated with the problems of the nuclear arms race and nuclear waste. There was a decline in funding of physics, too, as politicians tried to re-assert control over the priorities for science. What these criticisms of Big Science tell us is that there were alternatives, that large-scale science was not inevitable. But its very size and the huge investments made in its institutional basis, its self-perpetuating race to higher energies through trans-atlantic competition and its persuasive claims to be revealing the most fundamental aspects of our world all reinforced the cycle. Despite criticisms like Anderson's, every time a major new round of accelerator building was being discussed in the 1970s and 1980s the physicists got their money. In Europe, CERN funding was largely a political and diplomatic issue between its member nations. In the US, funding for high-energy physics became increasingly disconnected from military sources, being funded in the 1980s by the AEC's successor, the Department of Energy (DoE). Yet the military and ideological origins of the large accelerator programmes remained embedded within them. This was to be vividly illustrated in the early 1990s after the end of the Cold War. In 1993 the clouds that periodically appeared over high-energy physics funding turned into a storm over a construction site in Texas.

# CHAPTER 11

# DEATH IN TEXAS: THE END OF MEGASCIENCE?

By the early 1980s, the quest to explore the Standard Model and find the W and Z particles was well underway. From 1978–81 at CERN, the development of a technique known as 'stochastic cooling' allowed substantial improvement of beam intensity and quality. The technique was used to convert the SPS into a proton-antiproton collider, using an Antiproton Accumulator ring (AA). Two large experiments, designated UA1 and UA2, were devised to study the particle collisions. Run by international collaborations involving laboratories from several countries, as we have seen, UA1 and UA2 were competing not just with laboratories in the USA but also with each other. The size and complexity of the equipment – UA1's calorimeters (energy detection devices surrounding the central detection chamber) alone were each the size of a small school bus – and the numbers of people involved meant that the organisation required strong management. The energetic and outspoken Italian physicist Carlo Rubbia led the UA1 collaboration and was its official representative.

One member of the UA1 collaboration read a book on the Manhattan Project during the team's busiest and most stressful period, and was struck by the parallels between the two projects. As he put it: 'We had no time to have several designs. We had to have one design and get it through, whatever the costs. We had no way but just to succeed the first time.' Indeed, with fierce competition between groups working towards a particular goal, extensive decision-making by committee within a large organisation and real power concentrated in a few hands, there is more than a passing resemblance. Now, though, the race was not for a weapon but for the Nobel Prize. The first proton-antiproton collisions were seen at 270 GeV per beam in July 1981. In 1983, CERN announced the discovery of the W± and Z particles. The following year, Rubbia and Simon van der Meer (who developed the stochastic cooling technique) were awarded the Nobel Physics Prize for this work – not least as a result of a concerted campaign by CERN's management to gain recognition for the work of the laboratory as personified in Rubbia.

Meanwhile in 1981 CERN had decided on its next-generation accelerator: the Large Electron-Positron collider ring (LEP), planned to operate initially at 50 GeV per beam. The machine was commissioned in August 1989, and over the next five years four large detectors named ALEPH, DELPHI, L3 and OPAL observed more than 10 million Z decays using complex electronics and computers to reconstruct the colliding particles' behaviour. In 1994 the CERN Council approved construction of LEP's successor, the Large Hadron Collider (LHC), of which more later. And as operations at CERN became ever more complex, the US did not lag behind in the race to higher energies. In 1984 the Fermilab accelerator reached 800 GeV. A year later, the 900 GeV Tevatron at Fermilab began operation: it was later boosted to 1 TeV. These machines had taken longer than expected to come on stream, though, and US physicists were annoyed to have 'lost' the W and Z discoveries to CERN. In the early 1980s, American high-energy physicists had already presented plans for the next generation of accelerator. It was called the Super-conducting Super Collider, or SSC. At a planned cost of $4–6 billion to build and several hundred million dollars a year to operate, the idea was to accelerate two proton beams in opposite directions in a circular tunnel 52 miles around to an energy of 20 TeV, then collide them at 40 TeV – an acceleration energy 60 times greater than CERN's. The SSC would turn out to be 160 times larger than the biggest existing machine and require 16,553 acres of land to be acquired from over 700 landowners – shades of the Manhattan Project's Hanford and Oak Ridge sites.

The scientific rationale for the SSC was that it would enable physicists to explore problems in the Standard Model. It was expected that the SSC would reveal a new particle called the Higgs boson, which was expected to be found only at very high energies. Leon Lederman, director of Fermilab, explained it to a Senate hearing by asking them to think of a group of extraterrestrials watching a football game but unable to see the ball: 'They see a lot of people running around seemingly at random in a chaotic disorganized activity, but if someone postulated the existence of a soccer ball, then the whole thing becomes clear and simple and elegant.' Beyond finding the Higgs football, the Standard Model linked particle physics to cosmology, since accelerators could recreate energies and processes present near the beginning of the universe – the subject of Stephen Hawking's famous book *A Brief History of Time*. This has led some scientists to suggest that particle accelerators would be better called 'chronoscopes', since they can be seen as 'time machines that are using pieces of atoms to mimic the condition of the new-born Universe'.

As part of the campaign by American scientists to promote the SSC, Nobel Prize-winner Steven Weinberg produced a book of his own called *Dreams of a Final Theory*, in which he argued that the SSC was absolutely necessary to 'continue with the great intellectual adventure of discovering the final laws of nature'. 'The years since the 1970s', he argued, 'have been the most frustrating in the history of elementary particle physics. We are paying the price of our own success: theory has advanced so far that further progress will require the study of processes at energies beyond the reach of existing experimental facilities.' High-energy physics was 'stuck'. The SSC would kick-start it again. But there were other rationales for the new machine too. Some US physicists and their supporters felt humiliated at the loss of the W± and Z discoveries to Europe. The *New York Times* carried an editorial entitled 'Europe 3, US Not Even Z-Zero' and demanded that American physicists get 'earnest revenge'. For those who thought in such terms, the SSC would restore the US's pre-eminence in high-energy physics. Behind such thinking lay an overtly nationalistic agenda. As two senior US physicists and leading SSC promoters put it in 1985: 'If we forgo the opportunity that SSC offers for the 1990s, the loss will be not only to our science but also to the broader issue of national pride and technological self-confidence.' Tellingly, they added, 'When we were children, America did most things best. So it should again.'

Congress began to consider the issue in 1985. The project's supporters pointed out the various spin-offs from previous generations of accelerators, including radiation beams for medical therapy, for treating foods and materials and for etching integrated circuits. The SSC promised more economically beneficial spin-offs – the development of its superconducting magnets, for example, might well have applications in transportation or power generation, or in the further development of the powerful diagnostic technology of magnetic resonance imaging. The sheer scale of the project would be beneficial both to the national economy and to the region where it would be built. The tunnel would be filled with steel pipes and several thousand superconducting magnets energized by 19,400 km of superconducting cable. The super-conducting system would need 12 refrigeration plants with huge quantities of helium to cool the magnets. The two underground experimental halls comparable in size to a football stadium, each housing a particle detector weighing more than a battleship, would not just produce work for physicists, but a large number of industrial contracts and 'an awful lot of jobs' – including five to eight thousand in the area where it would be built. In the US concrete industry, the SSC project apparently came to be known as 'the big pour', and one commentator admitted that there were 'lots of physicists, politicians and businessmen out there lusting after it'.

There was strong support for the project from 1983 to 1989. The project was endorsed by the Reagan administration, as it increased spending on defence and science in an intensification of the Cold War (funding for the DoE Office of Energy Research grew 80% 1981–7, and Reagan's 1984 budget gave a \$6.9 billion increase to federal research and development funds). Running of the project was delegated to the Universities Research Association, the non-profit consortium of 69 universities that had been running Fermilab. As things got underway, with design work being undertaken at Lawrence Berkeley Laboratory, Brookhaven and Fermilab, more than half of the US states entered the site-selection competition. One magazine neatly called the SSC site-selection process 'quark barrel politics'. Eventually, in 1988, a site at Waxahachie in Texas was chosen. Some commentators have suggested that the powerful Texas congressional delegation and the fact that the then President-elect George Bush was from Texas were important factors in the decision. Certainly, the offer of 200 square miles of land and a $1 billion sweetener by the state of Texas cannot have hurt its chances.

Despite the fact that the final site-selection now removed the interest of most states from the project, the government voted funds for work to continue on the development: $225 million in 1989, and $450 million in 1991. By 1990, over 1,000 people were already working at Waxahachie. But opposition was beginning to grow, both inside and outside the physics community. Rustum Roy, a materials scientist and something of a maverick in US science, described high-energy physicists as 'spoiled brats' for wanting a huge new accelerator when the country was running up big deficits. Yet there was also serious opposition from less outspoken scientific critics from within the physics community – a 'loyal opposition' who admired high-energy physics but did not think the multi-billion-dollar cost of the project justified. Among such critics were distinguished physicists working in the fields of superconductivity and condensed matter. One of them was Philip Anderson, whom we have already met as a critic of Big Science in the 1960s. He criticised SSC promoters' claims 'that particle physics did everything from MRI and the computer revolution to the television screen and sliced bread', and argued that 'Dollar for dollar, we in condensed-matter physics have spun off a lot more billions than the particle physicists ... and we can honestly promise to continue to do so'.

Against this upsurge of opinion from within the physics community, there were increasing worries about the design of the accelerator and the possibility of escalating costs: estimates now projected the cost of the SSC at $4.4 billion, then $5.9 billion, then $8.25 billion, and finally at $11.8 billion. The DoE decided to sacrifice a major new project at Brookhaven to help finance the SSC, and doubts were voiced about the management of the project. The spiralling cost of the new machine came at a time when US politicians were coming to terms with the cost of ending the Cold War. With the fall of the Berlin Wall in 1989 and the dissolution of the Soviet Union in 1991, the clouds started gathering for the military-industrial complex and Big Science. Other large national programmes – the Space Station and the Reagan administration's Strategic Defense Initiative – were each costing $2–3 billion per year by the early 1990s. This increased military spending was met by Congress's refusal to cut back elsewhere, so that the USA faced budget deficits of hundreds of billions of dollars in the early 1990s. Politicians began to get serious about cutting these debts in the early 1990s (it was the major issue of the 1992 US election). For the politicians and policy-makers, 'strategic' (i.e. economically beneficial) research now became significant: beating the Europeans to the Higgs particle did not seem so important. Already in 1992 physicists were discussing their future in the post Cold War climate. In one discussion, a chemist pointedly asked: 'I wonder how the physics community, once so clearly identified with the atomic bomb and weapons of all sorts, and still envied by the rest of science for its ability to plan ahead, will convince the people who do the actual funding that, with the Cold War over, it is necessary to continue supporting the community.' The residual links between big physics and the Cold War security state were coming home to roost.

In the summer of 1992, the SSC's Congressional critics moved to have the project cancelled, one of them commenting that, 'The SSC is committing suicide ... by failing to be honest with the taxpayers as to how much it will cost.' Another doubted 'that the most pressing issues facing the Nation include an insufficient understanding of the origins of the universe, a deteriorating standard of living for high-energy physicists, or declining American competitiveness in the race to find elusive subatomic particles'. Despite the fact that some politicians saw the SSC as a way to keep scientists and engineers employed as the US evolved its post-Cold War political and economic order, the US Congress voted 232 to 181 to terminate the SSC project in June 1992.

The decision was reversed by the Senate, but now, disconnected from national security issues and competing with the Space Station project and budget cutbacks, the case against the SSC hardened. Though the Space Station cost more than twice as much as the SSC, there was substantial foreign investment in the project, the support of the aerospace industry and the promise of 75,000 jobs likely to be spread across the USA. Its proponents continued to claim that cancellation of the SSC would signal that 'curiosity-driven science is somehow frivolous and a luxury we can no longer afford', and that its cancellation would represent 'the killing of support for basic science in this country'. Under the new Clinton administration in June 1993, Congress again voted, this time by 280 to 150, to kill the SSC. Though the Senate again reversed the decision, the project was finally cancelled by a Congress vote of 282 to 143 against in October 1993.

Writing as decisions about the SSC were being made in 1993, Steven Weinberg suggested that competing big-money projects like the Space Station 'are too well protected politically by a network of aerospace and defense companies to be scrapped, leaving the SSC as a conveniently vulnerable target for a purely symbolic act of deficit reduction'. He also stated that, 'It may be that the closing years of the twentieth century will see the epochal search for the foundations of physical science come to a stop, perhaps only to be resumed many years later.' On both counts he was probably right. The cancellation of the SSC certainly made clear the close connection between Big Science, politics and the Cold War, at least in the USA. But it equally reflected the significance of opposition from physicists themselves. One physicist called the SSC 'perhaps the most divisive issue ever to confront the physics community'. Its supporters saw the decision as a 'senseless killing'; its opponents as a long-overdue 'comeuppance for high-energy physics'. Whichever way one sees it, the demise of the SSC was more a redirection of the relationship between science and government in the post-Cold War political and economic context than a repudiation of basic science itself.

The SSC case makes an interesting comparison with another flagship Big Science project on the books at the same time. The Space Station project survived the political process (albeit by one vote) just the day before the SSC collapsed. The Space Station was in fact expecting much greater cost overruns (up to $30 billion) than the SSC. But because it was not centred in a particular state, because contracts and jobs were more evenly distributed across the US and because NASA had significant foreign financial contributions (and better spin doctors), the Space Station was retained as a national project while the SSC was cancelled. Whatever the politics, with the cancellation of the SSC, many American physicists had to look elsewhere to continue their work. The only comparable institution is, of course, CERN. CERN too had been affected by the end of the Cold War, though the European experience was very different to that of the US. Finland and Poland joined CERN in 1991, Hungary in 1992, the Czech and Slovak Republics in 1993 and Bulgaria in 1999, bringing the number of member states to twenty. In 1995 Japan became a CERN Observer State. Following the SSC debacle, US physicists decided to participate in CERN's Large Hadron Collider (LHC) project. In 1997 the DoE recommended $500 million as a US contribution to the LHC, and the US also became a CERN Observer State.

The LHC at CERN is now the largest and perhaps the last venture in this kind of Big Science. In 1996 the energy of LEP was doubled to 90 GeV per beam in a machine designated LEPII. But with physicists believing that significant new discoveries are to be made around 1 TeV, they set a goal for the biggest machine yet seen: the LHC. It will use the existing 27-km LEP tunnel, 8,000 superconducting magnets and other elements of CERN infrastructure to achieve proton beam energies of 7 TeV. The detecting systems will be as large as office blocks, but will produce the tiniest particles yet known. They will produce as much data in one day as LEP's detectors do in a year and will handle as much information as the entire European telecommunications network (Figures 18 and 19).

The fate of the SSC and big physics presents an interesting contrast with another recent manifest-ation of Big Science: the Human Genome Project (HGP). Like the SSC, the idea for the HGP originated in

*Figure 18.* One of the huge, complex detectors being built by engineers and physicists for the Large Hadron Collider at CERN.

the early 1980s, when biological scientists began to discuss a large-scale project to map the complete human genome sequence. Early proponents envisioned several thousand people working on a crash programme at an estimated cost of some $3 billion. The proposal was taken up by the Office of Health and Environment in the US DoE, and they allocated $4.5 million to setting up a major initiative in 1987. The Department already supported work on the biological effects of radiation at Los Alamos, and as the major funder of US particle accelerators was quite used to supporting large-scale goal-directed projects. Now, it also saw the genome initiative as a way of revitalising

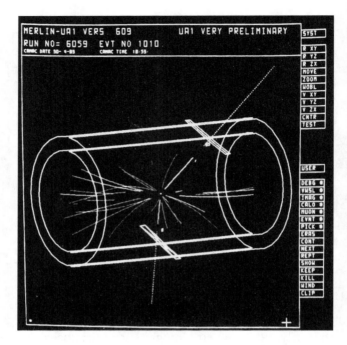

*Figure 19.* The complexity of Big Science: the huge detectors at modern accelerator facilities produce many millions of particle tracks, from which signature 'events' are identified by computers. Out of pictures like this the W and Z particles were discovered.

its national laboratories, whose bomb-making activities were in less demand as the Cold War wound down. In 1986–7 the DoE therefore established a five-year plan for a human genome mapping programme, including the development of automated sequencing technology and research into the computer analysis of data. Human genome research centres were set up at Los Alamos, Livermore and Lawrence Berkeley National Laboratories. As with the SSC, US international competitiveness was an important argument for genome advocates, especially industrialists, who pointed out that Europe and Japan seemed to be moving heavily into molecular biology and biotechnology.

Despite the enthusiasm of DoE, many biomedical scientists began to worry about the importation of what they saw as Big Science methods – big money, centralisation, bureaucracy, goal-directed projects – into traditionally small-scale, researcher-led science. One denounced the project as 'a scheme for unemployed bomb-makers'. Like the physicists at Fermilab and Brookhaven, they were concerned that large-scale funding for one project would bite into smaller-scale support for many others. Opponents also worried that a complete mapping of the human genome would be a complete waste of time – even perhaps bad science – since only about 5% of base pairs in human DNA code for genes. The National Institutes of Health (NIH) – chief supporters of biomedical research in the US – now began to take measures to wrest the genome project from the DoE, and succeeded in obtaining significantly more funding than DoE in 1987. In 1988 a panel of senior biomedical scientists backed the human genome initiative, but proposed a programme spread over fifteen years, and funded at $2 million per year. Part of this money was to be devoted to the development of technologies that would make sequencing rapid and cheap – so that the work could be spread more widely. Up to ten large multi-disciplinary laboratories around the USA were envisaged.

In 1988, NIH appointed James Watson to head a new Office of Genome Research (later the National Center for Human Genome Research). The project was divided between eight labs supported by NIH (responsible for mapping) and three DoE national labs (responsible for the development of sequencing technologies). Its budget grew rapidly from $39 million in 1989 to £135 million in 1991, when it became a formal US federal programme. Biomedical scientists continued to criticise the project, however, again on the grounds that it squeezed the rest of biology and subjugated it to Big Science. According to one critic: 'The human genome project is bad science, it's unthought-out science, it's hyped science.' Hyped it certainly was. Scientific journals and newspapers (as well as scientific critics) have tended to identify the HGP with Big Science – but did not do so with the AIDS research programme, which had a much greater annual budget. By now, the HGP was being lumped together with the SSC (and the Space Station) as big projects that were squeezing little science. But it was clearly a different *kind* of Big Science than the SSC. Though it dispensed big money, it did so in smallish doses, rather than spending it on one large facility. In effect it continued to support the small groups, loose coordination, academic freedom and institutional diversity that had always been characteristic of NIH funding in the biomedical sciences. Thus the HGP 'was the kind of Big Science – pluralist, practical, and payoff rich, unlike the SSC – that the political system remained willing to supply with the public resources that it requires'.

Of course, by the mid-1990s the HGP had become a race for results, not just between US and European laboratories, but also between public and private initiatives. Privately funded organisations such as the Institute for Genomic Research and companies like Celera Genomics set themselves up in competition with the publicly funded project, aiming to capture lucrative patents and to exploit genome information. Using 300 off-the-assembly-line sequencing machines and one of the world's fastest supercomputers, Celera Genomics adopted a 'shotgun' approach. Their aim was to sequence the entire human genome – all 3 billion bases – in three years. In a test of this 'shotgun' strategy (reminiscent in style of the crash implosion programme at Los Alamos in 1944), Celera announced its intention to sequence the genome of the fruit fly. By March 2000 the trial had worked, proving the success of the approach. By this time, in response to Celera's forcing of the pace, the public initiative had announced new goals: the complete sequence by 2003, with a 90% 'rough draft' in 2001 – a strategic move to undercut Celera and make information publicly available. The race was now on between Celera and the publicly funded programme. It was conducted not just in the laboratories, but also in the media through the hype of claim and counter-claim. The controversy continued up to and beyond the announcement of the draft human genome in 2001.

It is ironic that both the Celera project, with its Los Alamos-like 'shotgun' approach, and the public initiative, with its coordinated distribution of a single, goal-directed programme over many laboratories, share elements of Manhattan Project-style science. They also indicate the major shift of emphasis from the physical to the biological sciences at the end of the twentieth century – a trend mirrored in funding patterns and student recruitment alike. More to the point, however, with the death of the SSC and the imminent completion of the HGP, what lies ahead for Big Science?

## · CHAPTER 12 ·

# CONCLUSION:
# THE MYTHS OF BIG SCIENCE

To assess the future prospects of Big Science, we need to appreciate its past. The process by which Big Science came to prominence in the twentieth century was a long and complex one. Big Science did not spring fully formed into being after the Manhattan Project. Indeed, Big Science itself was not inevitable, driving all before it. At various stages during the course of the last century, scientists and others criticised the growth of science on a scale requiring the massive mobilisation of labour and resources. And, as we have seen, large-scale science has taken different forms in different places at different times. From the big telescopes of astronomers over the last several centuries to the scientific machine builders of the late nineteenth century and the ever-larger institutions of the twentieth, there seems to have been large, expensive equipment as long as there has been organised science. In physics, it is true, the period after World War Two saw a massive shift in the scale of some kinds of physics. Yet this was, in many ways, a continuation of trends that had begun well before the war. The real change after the war was not so much in the emergence of Big Science as in the consolidation of links between science (especially physics) and the national security state. Thus the story of Big Science is in important ways also the story of the relationship between Science and the State.

We have focussed in this book on high-energy physics as the iconic form of Big Science. But other kinds of Big Science have also been very important. Oceanography and space science are two good examples. Astronomy also continued to be at the forefront of Big Science. At an estimated cost of some £300 million, the European Southern Observatory's planned Atacama Large Millimetre Array with its 64 12-metre radio telescopes will be the largest ground-based astronomy project ever. And the Hubble Space Telescope, a $2 billion project involving NASA and the European Space Agency, has extended Big Science into space-based astronomy. The military too, of course, has continued to be an important sponsor of Big Science. The various national nuclear weapons and nuclear energy projects – the direct heirs of the Manhattan Project – have been perhaps the most significant examples of military Big Science, but missile, aircraft and reconnaissance satellite programmes have also been hugely important. The successors to the radar air-defence schemes of World War Two and the Cold War have been missile defence projects: in the 1980s, research on Ronald Reagan's Strategic Defense

Initiative ('Star Wars') cost $60 billion, and today, George W. Bush's National Missile Defense scheme, whose costs are estimated at anything between $60 billion and $200 billion.

Nor has all Big Science been successful science. Since the 1940s, physicists have been trying to achieve controlled thermonuclear fusion – the harnessing of the awesome power of the H-bomb to produce energy. Yet despite fifty years of research, billions of dollars and the creation of large-scale national and international projects such as the USA's Project Matterhorn, the Joint European Torus (a torus is a device for creating the plasmas in which scientists hope to produce and control fusion) and now ITER – a collaboration of engineers and scientists from Canada, Europe, Japan and Russia – the route to the successful exploitation of fusion has yet to be found. There were spectacular failures on the military side too: between 1946 and 1961 the US military spent more than $7 billion on its Aircraft Nuclear Propulsion programme – an attempt to produce an atomic-powered aircraft whose range would be so great that it was joked that it would have to land occasionally for the crew to re-enlist.

If Big Science was neither monolithic nor always successful, its various incarnations nevertheless did share certain key characteristics, many of them consolidated or developed in the Manhattan Project. Several features stand out. First, naturally, the *goal-directed* nature of the work. The imperative to design and build nuclear weapons before the enemy was what pushed the Manhattan Project forward with such speed. Central planning and coordination were the keys to achieving the Project's mission. The division of functions between the three main sites – Oak Ridge, Hanford and Los Alamos – served to maximise the efficiency with which each of the principal scientific and technical tasks was carried out. Within each site, a rigidly hierarchical organisation of labour both provided for the efficient allocation of work and provided a mechanism for ensuring some control of information flow – the so-called policy of 'compartmentalisation'. At the apex of the organisation chart was General Groves. Ambitious and driving, he brought his invaluable practical experience of large-scale military projects, pushed for resources whenever necessary and masterminded almost every aspect of the Project's organisation. Ernest Lawrence and Carlo Rubbia are his heirs.

Second, the crucial role of *industry*. At Oak Ridge and Hanford, crucial for the production of fissile materials for the weapons, work was on an industrial scale, and workers' experiences were not unlike those on the assembly lines of Henry Ford's automobile factories (indeed, the Manhattan Project as a whole was as large as the American automobile industry). Commercial giants like Du Pont, Union Carbide, Monsanto and the huge industrial base of the USA were key to the success of the Project, and could be drawn upon at very short notice in a war economy.

Even at Los Alamos, with its more academic atmosphere, industrial models of research organisation were important. Close relations with industry, the use of robust empirical solutions to problems and a 'don't say no' attitude prevailed. Under Groves' and Oppenheimer's direction, this attitude flourished on the Hill, too. Los Alamos also drew on practice in pre-war US industrial laboratories, like those of American Telephone and Telegraph, General Electric and Du Pont, but applied it on a much larger scale. Researches in these science-based companies tended to be broadly academic in style but often used hierarchical multi-disciplinary teams on projects based on the properties and behaviour of materials with potential applications. This then leads to a third key point: the interdisciplinary nature of work at Los Alamos. Because of the lab's mission, its research was closely linked to the behaviour of particular materials (e.g. plutonium) and technologies (e.g. B-29 bomb bays); the product had to be completely reliable, so that equipment was often over-designed to avoid risk. This conservative approach (as in the Christy gadget) applied to the organisation of work too. Multiple lines of investigation were pursued in parallel until specific advantages were found in one of them, then that line was followed through to rapid finalisation of design and production details. The work made extensive and systematic use of small-scale models before progression to larger-scale work, and there was close contact between theoreticians and experimentalists.

This framework permitted flexibility in the organisation of work, allowing, for example, the support of 'backup' or contingency programmes which were not necessarily high priority but which might at some time pay dividends within the overall goal of the laboratory. Despite the imperative to focus on the atomic bomb, for example, work was nevertheless supported on the 'super' or hydrogen bomb and on other developments like the 'implosion' technique, which offered the possibility of later payoffs in efficiency through more cost-effective use of the expensive fissile materials. The flexibility and utility of the interdisciplinary approach was best exemplified in the crash programme on implosion. That episode also showed the *interconnectedness* of the Project with other military and industrial programmes. Scientists with key skills could be requisitioned from almost anywhere to serve on the Hill. For example, the Project drew on the equally extensive radar programmes for instrumentation and personnel, ordnance and ballistics experts for the gun programme could be co-opted from industry or other military programmes, and, with the top-rated priority of the programme, industrial labs could also be roped in to provide resources, expertise and production facilities.

Looked at this way, the success of Los Alamos was as much an organisational as a scientific one, involving the coordination of lots of small-scale efforts and the integration of a wide variety of scientific skills into a new approach. Of course we have to be careful not to overstress the imposition of organisation. Sometimes, especially in the early days, the Project had a strongly *improvised* feel to it – in the recruitment stage Oppenheimer called it a 'somewhat Buck Rogers project'! At the same time, this organisational success came at a heavy price. The Manhattan Project as a whole was heavily bureaucratic in nature. With half a million employees it had a complex internal structure and economy, characteristic of large organisations, with the usual paperwork-shuffling, personnel, financial and other functions complicated by a large security apparatus. The upper levels of this bureaucracy were beset with personal and professional rivalries and empire building, and any number of committees and panels which made decisions about and constantly reviewed priorities, resource alloca-tions, targets, personnel and so on. This often led to problems such as lack of communication between institutions or groups within the project, or pirating of experienced or skilled people from one part of the project to another. Scientists have their share of prima donnas and egocentrics, like any other profession.

Whichever way one looks at it, the experience of science in World War Two was seared deep in the hearts and minds of the scientists who served in it. These experiences – not just in the Manhattan Project but in many other technoscientific programmes – fundamentally shaped the development of post-war science. For the younger and less experienced scien-tists, for whom war work was their first taste of real research, it naturalised a way of doing science which included large budgets, close cooperation between scientists and engineers, setting of targets and the hierarchical organisation of interdisciplinary teams. For the more experienced scientists, military hierarchy, planning and prioritisation added a new dimension to their thinking about what could be done in science. It was in the management of disparate teams and sometimes divergent personalities working under constant pressure that many of the leading scientists cut their organisational teeth: this was to be one of the important post-war legacies of the Project. The Los Alamos scientists became 'a kind of secular establishment – with the power to influence policy and obtain state resources largely on faith and with an enviable degree of freedom from political control'.

After the war, Los Alamos (which came in the public's mind to stand for the whole of the Manhattan Project) was often portrayed as a physicist's 'Shangri-la', where resources flowed freely and where everyone worked with determination in close collaboration towards the final goal of beating Hitler to nuclear weapons. Indeed, it has come to have mythic status: unsurprising, given the impact of Little Boy and Fat Man on world affairs. When physicists after the war turned to high-energy accelerators as a way of moving back towards a 'pure' science, they drew heavily only on their links with government and the military and the prestige of the Manhattan Project to obtain funding for their new machines. Initially through the work of Lawrence and his Berkeley machine-builders, then through the politics of inter-laboratory competition, the national security state supported the development of high-energy physics in the USA. It is noteworthy, however, that large-scale science did not develop in the other sciences that played so prominent a role in the Manhattan Project – chemistry, metallurgy, engineering. Indeed, it was not typical even of physics: research in solid state physics, optics, acoustics and many other fields never approached the gigantism of high-energy physics.

In fact, though they only amounted to about 10% of the American physics community, the high-energy physicists tended to be more publicly prominent than their colleagues, and to be more effective participants in debates about science policy. One reason for this was their reading of history. Apparently useless nuclear research in the 1930s had led to the development of nuclear weapons. In the Cold War, support of particle and high-energy physics was insurance that if anything of significance for national security emerged from the accelerators, the US would have a significant lead over the Soviet Union. As Steven Weinberg argued: 'Even those who do not care about the discoveries made by ... physicists may reflect that the community of high-energy physicists represents a reservoir of scientific talent that has served our country well, from the Manhattan Project of the past to today's work on parallel programming for supercomputers.' Nevertheless, as one commentator pointed out, the decision to cancel the SSC was a salutary lesson to those who thought that science started with World War Two 'as if there was no science in the world or in America' before then.

In the early Cold War, the politics of European integration and of the Atlantic alliance created the conditions for the establishment of CERN. With its very different political underpinnings, no questions of national security were overtly involved at CERN. Yet the new laboratory effectively entered a competition with the large American laboratories, beginning a transatlantic race to ever-higher energies that would last for half a century. As the machines got bigger and costlier, the particles got smaller and the theories more complex in a self-perpetuating cycle. The very different origins of the US national accelerator laboratories and CERN remind us once again that post-war Big Science was not inevitable, and that political factors were crucial to the way the large institutions took shape and interacted. CERN itself became the prototype for several other European collaborations – in space research, astronomy and molecular biology. But the different ways of working at US labs and at CERN – in the USA, for example, physicists and engineers had a much closer relationship than they did in Europe – demonstrates again that there were different *styles* of Big Science in different places.

Big Science has also come to mean something more than the science itself. By the 1980s the competition between the various accelerator laboratories and even between groups within laboratories like CERN had become in some ways more important than the physics. The guiding philosophy was something like 'build a bigger machine, get to energies that no other machine has reached, and you'll find something new'. Competition for Nobel Prizes as well as laboratory and national prestige fostered egotism, ambition and 'intellectual exhibitionism', just as in any other sphere of human activity. And of course awards and prestige meant access to resources and the means to build bigger and better experiments and accelerators. Carlo Rubbia put it like this: 'From [Ernest] Lawrence onward, if you really wanted to become famous in our field, you had to get yourself the biggest and most powerful accelerator.' 'Pure' physics, if there had ever been such a thing, was now a business. Yet in contrast to the entrepreneurial high-flyers and politicians, work for most of the physicists could be mundane. As one accelerator physicist said, 'You spend most of your working time not doing physics. You're learning whether this or that works, going through the hadron calorimeter looking for light leaks, going through the electronics looking for dud chips or bits of solder that somebody has dropped between two wires that have shorted something out. There's a lot of slum.'

As this paradoxical image suggests, Big Science in the twentieth century was not homogeneous. And now, at the start of the twenty-first, it may be that we are at the beginning of the end of Big Science. With the cancellation of the SSC, the Waxahachie site of the SSC itself was returned to the state of Texas, and the researchers who had been working there sought jobs elsewhere as Big Science suddenly contracted. In the wake of the terrorist attacks on the USA in September 2001, part of the former SSC site is rumoured soon to become home to a counter-terrorism training camp run by a Texas firearms training company. Though the fifteen or so miles of tunnels excavated had been filled in, they will be re-excavated and used to practice fighting in cave networks – well hidden from passing spy satellites. Meanwhile there are reports that Fermilab physicists are again discussing the possibility of a huge new machine. With the LHC about to come on stream at CERN, physicists are wondering if it will be powerful enough to flush out the Higgs boson. In order to avoid overlap with the LHC, US physicists are now talking about a large linear electron collider – possibly up to 20 miles long and costing $10 billion or more. And they, too, think that 9-11 may help: one of them hopes that the US will 'be more serious now and more willing to do complicated, expensive things – including big physics – for the general good. I think it's time this country starts worrying about its legacy'.

In the US, the future of Big Science is politics. On the other side of the Atlantic, too, CERN is now being touted as much for having invented the World Wide Web as for any of its achievements in high-energy physics. First proposed by Tim Berners-Lee and Robert

Cailliau in 1990 as a vehicle for managing and distributing the huge amounts of information generated at CERN, the WWW is indeed perhaps the only spin-off with significant social impact to have come out of fifty years of particle physics! The criticisms that have regularly been made of large-scale science – indeed the coining of the term 'Big Science' in the 1960s to capture its pathological character – seem no less valid today. The dissolution of the links between physics and the military-industrial complex after the end of the Cold War seems to suggest that Big Science, at least in physics, has come to the end of the road. Yet the National Missile Defense project certainly indicates the continuing viability of large-scale military research and development, while the Human Genome Project suggests that other forms of Big Science may still be viable, with commercial interests, rather than the military, as the driving force. Big Science, in one form or another, looks likely to be with us for some time to come.

# FURTHER READING

## 1 Introduction: Big Science and the Bomb

A good, broad overview of the growth of physics in the twentieth century and a selection of further references can be found in H. Kragh, *Quantum Generations: A History of Physics in the Twentieth Century* (Princeton University Press, 1999). A comprehensive collection of more specialist essays is J. Krige and D. Pestre (eds.), *Science in the Twentieth Century* (Harwood Academic Publishers, 1997).

An entertaining account of the race for the W and Z particles is given by G. Taubes, *Nobel Dreams: Power, Deceit and the Ultimate Experiment* (Random House, 1986). G. Fraser, *The Quark Machines: How Europe Fought the Particle Physics War* (Institute of Physics Publishing, 1997) is another readable account of the historical development of big physics. D. de Solla Price, *Little Science, Big Science* (Columbia University Press, 1963) focusses specifically on large-scale science and its longer-term historical development.

## 2 Long Before the Bomb: The Origins of Big Science

Price, *Little Science, Big Science*, is again the starting point. On big astronomy, see W.G. Scaife, *From Galaxies to Turbines: Science, Technology and the Parsons Family* (Institute of Physics Publishing, 2000). Further details on late nineteenth- and early twentieth-century machine physics and links between universities and industry can be found in M. Eckert and H. Schubert, *Crystals, Electrons, Transistors: From Scholar's Study to Industrial Research* (American Institute of Physics, 1990). For the early history of radioactivity, see M. Bunge and W.R. Shea (eds.), *Rutherford and Physics at the Turn of the Century* (Science History Publications, 1979).

## 3 Science, the Military and Industry: The Great War and After

On military science in the Great War, G. Hartcup, *The War of Invention: Scientific Developments, 1914–18* (Brassey's, 1988) is very accessible.

Though rather uncritical, J.G. Crowther's *The Cavendish Laboratory, 1874–1974* (Macmillan, 1974) gives a good overview of the Cavendish. The emergence of nuclear physics and the early development of particle accelerators are thoroughly described in J.L. Heilbron and R.W. Seidel, *Lawrence and his Laboratory: A History of the Lawrence Berkeley Laboratory* (University of California Press, 1989).

R. Buderi, *The Invention that Changed the World* (Simon & Schuster, 1996) is a good general history of radar. The organisation of British radar is vividly captured in D. Zimmermann, *Britain's Shield: Radar and the Defeat of the Luftwaffe* (Sutton, 2001).

## 4 From Fission to Mission: The Origins of the Manhattan Project

The scientific background to the Manhattan Project, the project itself and the personalities of the various people involved are thoroughly covered in R. Rhodes, *The Making of the Atomic Bomb* (Simon & Schuster, 1987). Rhodes also provides an extensive range of further references.

The bureaucratic organisation of the Manhattan Project is brilliantly described in P. Hales, *Atomic Spaces: Living on the Manhattan Project* (University of Illinois Press, 1997). A rich source of images of the various Manhattan Project sites is: R. Fermi and E. Samra, *Picturing the Bomb: Photographs from the Secret World of the Manhattan Project* (Harry N. Abrams, 1995).

## 5 Los Alamos: Little Science on a Big Scale?

Rhodes, *The Making of the Atomic Bomb*, is again a good starting place. My account of the work at Los Alamos draws heavily on the detailed history in L. Hoddeson, P.W. Henriksen, R.A. Meade and C. Westfall, *Critical*

*Assembly: A Technical History of Los Alamos During the Oppenheimer Years, 1943–1945* (Cambridge University Press, 1993).

Robert Serber's Los Alamos lectures are available in: R. Serber, *The Los Alamos Primer: The First Lectures on How to Build an Atomic Bomb* (University of California Press, 1992). More information on the British involvement in the Manhattan Project can be found in F.M. Szasz, *British Scientists and the Manhattan Project: The Los Alamos Years* (Macmillan, 1992).


## 6  Thin Man Becomes Fat Man: The Plutonium Implosion Programme

Again, Rhodes, *The Making of the Atomic Bomb*, gives the most readable account, while Hoddeson *et al.*, *Critical Assembly*, provides a much more detailed picture.


## 7  From Trinity to Victory: Making and Using the First Nuclear Weapons

As well as Hoddeson *et al.*, *Critical Assembly*, the Trinity test is described in detail in F.M. Szasz, *The Day the Sun Rose Twice: The Story of the Trinity Site Nuclear Explosion, 16 July 1945* (University of New Mexico Press, 1984).

There is a large literature (and substantial historical disagreement) on the controversial decision to drop the bomb. It is well summarised in J.S. Walker, *Prompt and Utter Destruction: Truman and the Decision to Use the Bombs Against Japan* (University of North Carolina Press, 1997).

## 8 After the Bomb: Big Science and National Security

On the relationship between physics and the post-war national security state, see S.W. Leslie, *The Cold War and American Science: The Military-Industrial Complex at Stanford and MIT* (Columbia University Press, 1993); D. Kevles, *The Physicists: The History of a Scientific Community in Modern America* (Harvard University Press, 1995, 2nd edition); and J. Wang, *American Science in an Age of Anxiety: Scientists, Anticommunism, and the Cold War* (University of North Carolina Press, 1999). The transformation of US universities in this period is described in R.L. Geiger, *Research and Relevant Knowledge: American Research Universities Since World War II* (Oxford University Press, 1993).

For a readable account of the development of the H-bomb, see R. Rhodes, *Dark Sun: The Making of the Hydrogen Bomb* (Simon & Schuster, 1995).

## 9 From Big Science to Megascience: The Age of the Accelerators

Fraser, *The Quark Machines,* gives an engaging account of post-war developments in accelerator physics, as does M. Riordan, *The Hunting of the Quark* (Simon & Schuster, 1987).

On Lawrence's post-war laboratory at Berkeley as well as contemporary songs (including the one in the text) and cartoons about the development of Big Science, see http://www.aip.org/history/lawrence. R.P. Crease, *Making Physics: A Biography of the Brookhaven National Laboratory, 1946–1972* (University of Chicago Press, 1999), gives the history of the Brookhaven laboratory.

A detailed treatment of the different kinds of detectors and the changes in material and organisational culture in big physics over the course of the century can be found in Peter Galison, *Image and Logic: A Material Culture of Microphysics* (University of Chicago Press, 1997).

On the Space Age and Big Science, see W.A. McDougall, … *The Heavens and the Earth: A Political History of the Space Age* (Basic Books, 1985).

## 10  The Invention of 'Big Science': Large-Scale Science as Pathological Science

Alvin Weinberg's original critique of Big Science is in 'Impact of Large-scale Science on the United States', *Science*, vol. 134 (21 July 1961), pp. 161–4. His further thoughts are in *Reflections on Big Science* (MIT Press, 1967).

Other contemporary views on which I have drawn are D. Greenberg, *The Politics of Pure Science*

(University of Chicago Press, [1969] 1999, 2nd edition) and H. Rose and S. Rose, *Science and Society* (Pelican Books, 1970).

## 11 Death in Texas: The End of Megascience?

On CERN, again see Taubes, *Nobel Dreams*. Despite the eloquent defence in Steven Weinberg, *Dreams of a Final Theory* (Vintage, 1993), the story of the cancellation of the SSC is engagingly told in Kevles, *The Physicists* (2nd edition). Details of its subsequent fate can be found on the website 'Super Collider Blues', http://abcnews.go.com/sections/Scitech/DailyNews/collider2_020430.html.

For the Human Genome Project, see D. Kevles, *The Code of Codes: Scientific and Social Issues in the Human Genome Project* (Harvard University Press, 1992).

## 12 Conclusion: The Myths of Big Science

On national nuclear programmes and missile defence schemes and other military projects, see http://www.fas.org/index.html.

CERN's role in the birth of the Internet is described in J. Naughton, *A Brief History of the Future: The Origins of the Internet* (Phoenix, 2000).

Further details on many of the themes developed in this book can be found in P. Galison and B. Hevly (eds.), *Big Science: The Growth of Large-scale Research* (Stanford University Press, 1992).